758 SS

D0893502

DATE D

THE RESTLESS UNIVERSE

THE RESTLESS UNIVERSE

Nigel Henbest and Heather Couper

Foreword by Professor A Boksenberg FRS
Director, Royal Greenwich Observatory

GEORGE PHILIP
London Melbourne Milwaukee

Illustration Acknowledgements
FRONT JACKET Anglo-Australian Observatory; BACK
JACKET NASA; 1 Royal Observatory, Edinburgh;
2 Science Photo Library; 3 NASA; 4 NASA;
5 NASA; 6 NASA; 7 NASA; 8 NASA; 9 NASA;
10 NASA; 11 NASA; 12 Martin Grossman;
13 NASA; 14 M. Maunder; 15 Royal Observatory,
Edinburgh; 16 Anglo-Australian Observatory;
17 NASA; 18 NASA; 19 Science Photo Library;
20 Royal Observatory, Edinburgh; 21 Space Fron-
tiers Limited; 22 Courtesy of Bell Laboratories;
23 National Astronomy and Ionosphere Center (op-
erated by Cornell University under contract with the
National Science Foundation); 24 Science Photo Library.

British Library Cataloguing in Publication Data
Henbest, Nigel
    The restless universe.
    1. Astronomy
    I. Title   II. Couper, Heather
    520   QB43.2
ISBN 0 540 01069 3

Printed in Great Britain by
Butler & Tanner Ltd
Frome and London

58,998

CONTENTS

FIGURES

COLOUR PHOTOGRAPHS

Foreword

The history of astronomy is a history of receding horizons. Every increase in telescopic power and every gain in the sensitivity of instruments has opened up new depths of space for exploration, and the centre of astronomical interest has tended to follow the limits of observation. Knowledge, too, has its horizons. Every science contains a core of well-established fact and, surrounding this, a dim nebulous region in which fact and hypothesis are mingled, where it is hard to distinguish truth from error – a region of conjecture. Here is the frontier of knowledge, the region in which progress is made.

Astronomy is not a subject apart but is an amazing application of physics and chemistry, which requires of its practitioners a fertile imagination and adventurous thinking. It is a subject truly needing the close involvement of man. Indeed, it is remarkable how well we fit into the Universe, and our very existence seems to demand such a critical balance in its properties and evolutionary progress as to defy mere chance: so much so that astronomers are led to the startling proposition that our Universe is the way it is *because* we are here to observe it, and further, that it is only one amongst a very large ensemble of other possible but mostly lifeless universes.

The Restless Universe takes us on a fascinating tour in space and time from the locally familiar to the very End of Knowledge; from the Solar System across the breadth of our Galaxy; then to other galaxies beyond, and on to the violent quasars which we can see spaced out towards the observable edge of the Universe. We are taken back earlier and earlier in the life of the Universe, to within a tiny fraction of a second from the time of the Big Bang, then forward again until we are weighing up the possibilities for the Universe's ultimate end. In all this it is fitting that the authors treat astronomy as a subject directly relevant to man, not as an abstruse intellectual discipline: it is, after all, *our* Universe.

Professor Alec Boksenberg
Royal Greenwich Observatory

— I —
Stardust

Dusk falls slowly over the Warrumbungle mountain range in New South Wales. Gum trees fade from sight, merging into gathering darkness, as the small country town of Coonabarabran shuts down for the night. But 30 kilometres (19 miles) away, at the top of a 1000-metre (3280-foot) peak, the day's work is just beginning.

Within a white-painted tower, 50 metres (164 feet) high, a 335-tonne monster of steel and glass begins to stir. As the dome above it splits open to reveal the night sky, the world's most sophisticated telescope latches on to the light coming from a faint, remote galaxy – half a universe away from us. Sensitive enough to see a candle on the Moon, accurate enough to point exactly at a target as small as a pinhead at a hundred paces under the guidance of its computer 'brain', the Anglo-Australian Telescope is one of the super-precision high-technology instruments that Man is now using to probe the Universe around us – the Universe of which we, and our planet Earth, are an insignificant part.

No longer do astronomers strain their eyes peering through long tele-scopes on the coldest of winter nights, charting the positions of stars and planets, and sketching their observations with numbed fingers. The Anglo-Australian Telescope is a dumpy structure, its light-collecting mirror 3.9 metres (13 feet) across; and it needs no human hand to touch it. The astronomers sit in an adjacent control room, warmed to 22 °C (71 °F), where they are confronted by panels of buttons, keyboards, flashing lights and TV screens. From here they instruct the computer that drives the telescope. A TV screen shows them the view through the telescope's giant eye, but it is there just as a check that all is working correctly. The light is detected and analysed, not by an astronomer's eye, but by serried arrays of electronic detectors, backed up by mini-computers, which take in raw starlight and break it up by wavelength, looking for the significant features that interest the astronomers. Eventually they automatically churn out all this information on another TV screen or on the long roll of paper emerging from a chattering line-printer.

The wealth of information pouring from observatories all over the world has shown astronomers what an unimportant position our planet Earth has in the wider scheme of the Universe. The Earth is one of nine planets circling the Sun, together comprising the solar system. The planets range in size from Pluto, only one-quarter Earth's diameter, up to Jupiter which is eleven times our planet's size. But all are relatively cool globes, relying on the Sun for light and heat. The Sun outweighs its planets a thousand times over; within its vast bulk nuclear reactions produce huge amounts of light and heat, and its immense gravity keeps the planets in orbit about it.

On a human scale, the solar system seems vast. It is 150 million kilometres (93 million miles) from the Sun to the Earth; and almost 6000 million kilometres (3750 million miles) out to the most distant planet, Pluto. But human ideas of distance are paltry on the cosmic scale. The nearest star, Proxima Centauri, lies 40 million million kilometres (25 million million miles) away – seven thousand times further than Pluto.

The kilometre is far too small a unit for dealing with such distances. Astronomers use instead the *light year*, the distance light travels in one year. A beam of light moves at the unimaginably high speed of 300,000 kilometres (186,000 miles) *per second*. A light beam could girdle the Earth seven times in one second, and reach the Moon in $1\frac{1}{4}$ seconds. Light from the Sun takes $8\frac{1}{3}$ minutes to reach the Earth, and $5\frac{1}{2}$ hours to distant frozen Pluto. It takes $4\frac{1}{3}$ years to reach Proxima Centauri.

The nearest star is thus about 4.3 light years away. One light year works out to 9.5 million million kilometres (6 million million miles), and it forms a very convenient ruler to the stars, and beyond. The bright star Sirius is 8.7 light years away; bright red Betelgeuse in Orion lies 650 light years from us. Most of the stars in the night sky are tens or hundreds of light years off, while the telescope reveals more distant stars, dimmed by distance. All these stars, however, are part of one huge star system: the Milky Way Galaxy.

The Milky Way consists of some hundred thousand million stars – of which the Sun is one very average member – arranged in a flattened disc. Within this disc, most stars are bunched together in two curved arms that spiral outwards from the centre (see Figure 18, p. 127). The Sun lies just on the inside edge of one of these spiral arms, 30,000 light years from the Galaxy's centre. The huge assemblage of stars measures some 100,000 light years from one side to the other.

Beyond our Milky Way Galaxy lie thousands of millions of other galaxies. Some of them are also spirals – including our near neighbour, the famous Andromeda Galaxy. Others are irregular in shape, while many are simply round balls of stars (Chapter 14). Once we are dealing with the distances to galaxies, the figures become even more staggering. Even the relatively nearby Andromeda Galaxy is $2\frac{1}{4}$ million light years away. Large telescopes like the Anglo-Australian Telescope now routinely investigate light from galaxies over a thousand million light years away across the Universe.

But such large instruments have revealed more than just the geography of the Universe – 'where' the planets, stars and galaxies are. They have opened an entirely new perspective on the Universe, revealing it to be a place of frantic activity, constant change, and of violent death and re-birth. Until a few decades ago, space beyond our planet Earth seemed to be an unchanging panorama. This feeling pervades us when we venture out under the black velvet canopy of a dark night sky. The stars seem constant from night to night, never changing in position, little – if at all – in brightness. They are individual hard jewels of light, all separate from us, and from one another. In front of the permanence of the stars, the brighter lights of the planets move staidly and predictably across the sky, as they follow their ordained orbits about the Sun.

All this now turns out to be illusion. The heavens are seething with activity. We are deceived because human life is too short, our mortal lives only a few decades. The stars too have lives – but their spans are measured in millions, or even thousands of millions, of years. This is the timescale that the Universe works on. During our human lifetimes, we see only a snapshot of the Universe. Even if we collect astronomers' observations back over the entire recorded history of man, we can piece together only a few frames of the epic movie of the Universe's history.

All the stars in the sky are moving – and at tremendous speeds. The stars nearest the Sun are milling about in all directions at some 100,000 kilometres (62,500 miles) per hour; and all these stars together circle the centre of our Galaxy at almost a million kilometres (625,000 miles) per hour. In huge gas clouds – called nebulae – new-born stars flare into life, and their powerful radiation sweeps away the remaining tatters of gas around them. Old stars destroy themselves in spectacular supernova explosions, for a brief period outshining a thousand million suns. And though the stars are well-separated from one another, they are by no means isolated. Filling the space between them is a tenuous gas, mainly hydrogen, so rarefied that an open matchbox in space would contain only half-a-dozen atoms. But the interstellar spaces are so vast that there is a tremendous amount of matter spread out here. Our Galaxy consists of a hundred thousand million stars; and the thin gas between the stars contains enough matter to make another ten thousand million stars.

This interstellar gas is churned by the stars' activity. From the surfaces of hot stars, 'winds' blow outwards and ruffle the gases; supernova explosions excavate emptier cavities in the already tenuous interstellar gas. In this ceaseless turbulence, denser pockets of gas become compact enough to hold themselves together by their own gravity. These become nebulae, and in them a new generation of stars is born.

Earth is not immune from this activity. Our solar system was born in a nebula compressed by a supernova explosion, some 4600 million years ago (Chapter 6). Subsequent supernovae must have bathed our planet in penetrating radiation, perhaps causing mutations in the evolving life-forms here. Some astronomers have suggested supernovae caused climatic changes on Earth, bringing on the great Ice Ages; others have argued that the Ice Ages have occurred when the Sun has carried the solar system through a dense, dark nebula. Either way, scientists now accept that extraterrestrial influences can mould the Earth's climate, and thus indirectly – as well as directly – affect life on our planet (Chapters 2 and 11).

Extraterrestrial dangers lurk nearer to home as well. Within the solar system there are large hunks of rock – the asteroids and comets – which pose an ever-present threat to the Earth (Chapter 6). A comparatively small piece hit our planet in 1908, and exploded with the force of a 12 megatonne H-bomb. Fortunately it landed in the wastes of Siberia, but it could easily have hit a densely-populated area like Europe, with catastrophic effects.

Our planet is exposed to all the influences of the Universe, subject to invisible currents and predators as it swims through the oceans of space. The Earth does, however, wear a very effective armour: the invisible cloak of its atmosphere. The air about us does more than provide oxygen for breathing; it protects us from many of the dangers from space. Although the occasional asteroid does penetrate the atmosphere, most of the smaller rocky grains burn up harmlessly 100 kilometres (62 miles) above our heads, shining briefly as 'shooting stars' (meteors).

The air also foils the onslaught of 'cosmic rays' – not in fact radiation, but subatomic particles shot out by supernova explosions. Although these particles are unimaginably light in weight – it would take a million million million million to weigh 1 gramme (.035 oz) – they travel so fast that each packs the punch of a tennis ball from a powerful serve. Cosmic rays are fortunately stopped when they hit molecules in the air, and dissipate their energy harmlessly.

A vast range of radiations from space also find our atmosphere an impenetrable barrier. Light is one form of *radiation*, but only a minor part of the entire spectrum. Electromagnetic radiation consists of 'waves' of electric and magnetic fields, travelling at 300,000 kilometres (186,000 miles) per second (the 'speed of light'), and different kinds are characterised simply by the wavelength – the distance from one 'crest' to the next 'crest' along the wave.

Our eyes are sensitive to radiations with wavelengths between 390 nanometres and 750 nanometres (where one nanometre is a millionth of a millimetre), and we call such radiation 'light'. Ordinary light is a mixture of all wavelengths, but they can be split up by various devices (like a prism) to spread out the wavelengths into the colours of the rainbow. The shortest wavelengths, around 390 to 450 nanometres, are seen as violet to blue shades. Intermediate wavelengths are green to yellow, and the longest waves, over 680 nanometres, appear red to the human eye. Light occupies only a small part of the entire spectrum, and it is unusual in being able to

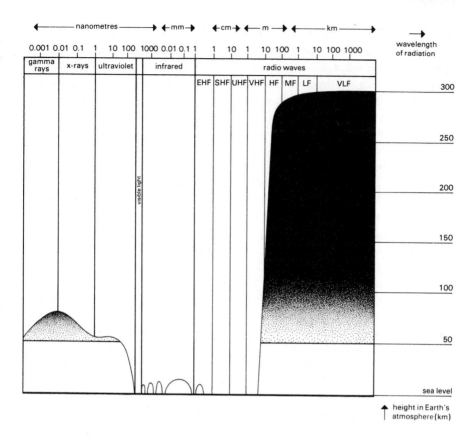

I THE ELECTROMAGNETIC SPECTRUM *The electromagnetic spectrum of radiation from space ranges in wavelength from the very short gamma rays (wavelengths marked in nanometres, where 1 nanometre is a millionth of a millimetre) to the longest radio waves which are over 1000 kilometres (620 miles) in wavelength. Ordinary visible light forms only a tiny part of the total spectrum. Visible light is, however, one of the only two wavelength 'windows' where radiation can penetrate the Earth's atmosphere down to sea level completely unimpeded; the other is short-wavelength radio waves. Infrared radiation is absorbed by water vapour and carbon dioxide in the lower atmosphere; ultraviolet and shorter wavelengths by ozone molecules and other atoms in the stratosphere; and the longest radio waves are reflected back into space by the highest layer in the atmosphere, the ionosphere.*

penetrate our atmosphere from space. Hence the Sun can light up Earth's surface, and our eyes can see the stars at night.

The longest wavelength radiations are radio waves, with wavelengths from 1 millimetre (.04 inch) upwards. The longest of the radio waves from space are reflected back by the uppermost layers of our atmosphere, the ionosphere. Radio waves can reach the ground, however, if their wavelengths lie between 1 centimetre (.40 inch) and 30 metres (98 feet). Apart from light, these are the only radiations from space that penetrate our protective atmosphere. Astronomers have taken full advantage of this fact by constructing radio telescopes which scoop up these cosmic radio waves.

Infrared radiation has wavelengths longer than red light, but shorter than most radio waves. It is commonly called heat radiation, because it is the radiation that fires give out. But although the warmth of a fire will radiate across a room, infrared is gradually absorbed by water vapour and carbon dioxide in the atmosphere. Infrared radiation from space is absorbed well before it reaches the ground.

Our atmosphere shields us from all the wavelengths shorter than light too. Ultraviolet rays are absorbed by a layer of ozone gas 50 kilometres (30 miles) up; the still-shorter X-rays and gamma rays are stopped when they hit molecules of air at even higher levels. We have every reason to be grateful for our atmosphere's shield, because these short-wavelength radiations are extremely damaging. The little ultraviolet that does filter down gives us a suntan – but a prolonged lie on a sunny beach will result in sunburn. Without the ozone layer, we would risk sunburn every time we stepped out of doors. X-rays penetrate deeper into the body, and excessive doses can damage internal organs. Particularly at risk are the reproductive cells. X-rays can easily alter the genes, and produce deformed children. Since the Sun produces copious amounts of X-rays as well as ultraviolet, our overhead shield is essential for life on Earth to survive at all.

This beneficial shield is, however, the bane of astronomers' lives. To find out about the Universe, we need to investigate all the radiations coming from 'up there', to expose our instruments and detectors to the full flood of radiations and subatomic particles coming from space. A modern optical telescope can wring the last drop of information out of the light coming from space, but light tells only part of the story. Many of the most exciting objects in the Universe do not emit ordinary light; to our eyes, and optical telescopes, they are dark and unseen in the blackness of space, but they may emit a pyrotechnic display of other radiations. Pulsars 'flash' regular pulses of radio waves across the Universe; the gas near black holes 'shines' briefly but gloriously in a burst of X-rays before it disappears into the hole (Chapters 9 and 10).

Modern astronomers are exploiting twentieth-century technology to the full, in their quest to absorb our cosmic environment and its full complement of messages from the farthest reaches of the Universe. Optical astronomy has made enormous strides by breaking with tradition – precision-engineered, computer-controlled telescopes with electronic light detectors are one product of this quest. But even an instrument like the Anglo-Australian Telescope – and its counterparts in the United States, Hawaii and Chile – is limited by our atmosphere. The air may be transparent to light, but moving air currents bend the light beams back and forward a little. To the romantic, this is essential to the romance of star-gazing, for it causes the gentle 'twinkling' of the stars. But it is a major headache for astronomy. 'Twinkling' means it is impossible to ever take a sharp photograph of any astronomical object from the Earth's surface. In the mid-1980s, astronomers will launch a telescope into space, to catch the light directly. This Space Telescope will orbit the Earth incessantly, carrying out observations twenty-four hours a day.

These optical astronomers will only be emulating what their less-fortunate colleagues have had to do for years. X-ray astronomers and ultraviolet astronomers cannot detect their radiations at all from Earth, and they must use detectors carried above the atmosphere in satellites. Remotely-controlled from laboratories on Earth, these unmanned 'telescopes' have been returning information on the whole range of radiations reaching us from the depths of space. The Einstein Observatory, which orbited the Earth, for example, carried an X-ray telescope that could photograph the X-ray sources in the sky and show as much detail as the largest optical telescopes reveal when photographing light from an astronomical object. It was only fifteen years between the first detection of a cosmic X-ray source (by a brief rocket flight) to the launch of the Einstein Observatory; and the improvement in X-ray telescopes in that time was far greater than the improvement in optical telescopes between Galileo's first crude 'optick tube' of 1609 and a modern optical telescope like the Anglo-Australian Telescope.

Man is now consciously reaching beyond his sheltering atmosphere to experience the ebb and flow of radiation and particles all about our spaceship Earth, to sample the cosmic ocean. As far as our immediate neighbours in the solar system are concerned, our contact is even more direct. Men have landed on the Moon: they have kangaroo-hopped over it, driven across it, drilled its surface to disinter its secrets. Unmanned spacecraft have landed on neighbouring planets Mars and Venus, others have flown past Mercury, Jupiter and Saturn, returning beautifully detailed photographs of the planets and their moons. They have 'sniffed' the planets' environs to detect what atoms and particles surround them, and extended feelers to measure their magnetic fields.

Astronomers now see the other planets as worlds which can be studied like our world: they are studying the 'geology' of Venus and Mars; the 'meteorology' of Jupiter and Saturn; the 'vulcanology' of Jupiter's moon Io, with its seven huge active volcanoes. The Earth is put into its context as one planet amongst a whole family; each of them as unique and interesting in its own way as ours.

To put our place in perspective, we can say that Earth is a small planet, circling an average star, in an undistinguished neighbourhood of an ordinary kind of galaxy. We are, however, an integral part of the Universe. Without stars like the Sun, there would be no galaxies. Earth was born as a companion of the Sun. We interact with the flux of radiations and particles that surround us in space. The very atoms that make our world

- and our bodies - came from space around, from the astronomical events that shape our Universe.

All matter is made up of a few *fundamental particles*: protons, neutrons and electrons. *Protons* and *neutrons* are roughly the same weight, and differ mainly in the fact that protons carry a positive electric charge. These two kinds of particle can stick together to build up dense agglomerations called *nuclei*. An *electron* is a much less massive particle, only 1/1800 as heavy as a proton or neutron, and it carries a negative electric charge. Out in space, most nuclei and electrons float around independently, in a kind of electric soup; but in the more sheltered parts of the Universe (like the interior and surfaces of the planets), electrons slip into stable orbits around the nuclei to form *atoms*. The atoms of any particular chemical element are distinguished by a particular number of protons in the nucleus, and an equal number of electrons in orbit around - hydrogen has just one of each, for example, while carbon has six and an iron atom twenty-six.

Hydrogen atoms in water were created in the Big Bang, the explosion which gave birth to our Universe. The carbon atoms in our body cells were built up from hydrogen in the bodies of stars, and puffed out into space. The iron atoms which colour our blood red were forged in the cores of massive stars, and blasted out in supernova explosions. Mixed up in the vast reaches of interstellar gas, these atoms ended up in a nebula; and here the solar system was fashioned by the inexorable force of gravity.

We are brought up from infancy to look at our 'environment' as the land and sea of our world; the blue sky and the clouds, the plants and living creatures that populate our nursery rhymes and story books. Astronomy today tells us to broaden our view. Our planet is part of a larger environment. The skies are, in reality, dark space, and the clouds in it are the glowing nebulae of star birth. Strange inanimate monsters lurk: rapidly flashing pulsars, the unpredictable explosions of supernovae, the irresistible maws of black holes.

Strange, powerful and frightening it may be, but we are a part of this Universe. It fashioned us, in as true a sense as evolution fashioned us on this planet. An extraterrestrial perspective demonstrates that - by cosmic standards - the Earth truly is the Garden of Eden. It brings home just how important it is to preserve our unique corner of the Universe - if only because it is only here (so far as we know) that intelligence has arisen, intelligence that can transcend its own physical limitations and encompass in its mind the entire Universe.

— 2 —

Planet Earth

Look at the solar system from outside, and there's nothing at all special about the third planet out from the Sun – the planet we call the Earth. From a passing spaceship, we would notice Jupiter, the largest of the planets, and undoubtedly our gaze would be drawn by the beautiful rings of Saturn. We might be lucky enough to see a comet, with a shining tail crossing millions of kilometres of space. Planet Earth cannot rate with the top tourist spots for a casual visitor to the solar system.

Earth is an intermediate-sized world. It is outclassed by the four outer giant planets: Jupiter is eleven times Earth's size, Saturn slightly less, while even the smaller giants, Uranus and Neptune, are as wide as four Earths. On the other hand, the other planets are rather smaller than Earth. Venus is just short of the Earth's size, but Earth is twice the size of Mars, three times Mercury's diameter, and four times the size of tiny Pluto, which orbits the Sun out beyond the giant planets.

What would first strike an extra-terrestrial visitor about the third planet is not Earth itself, but our orbiting Moon. Most of the planets have natural satellites, but almost all of these moons are comparatively small – less than a tenth the planet's own size. Our Moon is fully one-quarter the diameter of Earth. From afar, the Earth and Moon would appear like a double planet, rather than a planet dominating an insignificant satellite. Yet even here Earth is not unique. Pluto too is a double planet: its newly-discovered satellite Charon is about half Pluto's own size.

It's only when we get close to Earth that we realise just what an unusual world it is. The third planet is the only member of the Sun's family to have liquid water on its surface – seas and oceans, rivers and lakes. It's the only planet whose atmosphere contains the very chemically reactive gas oxygen. And Earth is the only world we know that harbours life. The range of habitats in which life flourishes is immense, from the depths of the seas to the mountain peaks, from torrid desert to frozen poles; and its variety is as overwhelming. Both plants and animals range from microscopic plankton cells up to individuals weighing tonnes – whales and giant redwood trees.

As inhabitants of Earth, we take this description for granted. Early astronomers assumed that the other planets would have habitats much like ours, and intelligent beings would be living there. Even as recently as the eighteenth century, the great astronomer William Herschel spoke confidently of creatures on all the other planets – and even residing on the Sun, under its fiery surface. In the nineteenth century, many astronomers believed there was intelligent life on Mars, and recommended setting out huge patterns in the Sahara Desert to attract their attention.

Now we know that Earth is the odd planet out – the only place in the solar system where life *can* survive. The other planets differ amongst themselves as much as they differ from the Earth, but they all share one characteristic: they are lifeless. The study of other planets has made us see Earth more in context, as a planet amongst planets, and to appreciate its unique attributes which make life possible. One aim of planetary scientists is to explain why Earth has developed this way.

On the other hand, we can see Earth's watery surface, its oxygen-rich atmosphere and the panoply of life as a thin veneer on a planet that shares many characteristics with the others. Scientists know more about the Earth than about any other planet. Geologists have investigated the rocks of Earth's crust, while their colleagues the geophysicists have indirectly probed our planet's deep interior. Oceanographers have studied the depths of the oceans, and the water flow; meteorologists have investigated the flow of gases in our atmosphere.

To begin with, the 'earth sciences' were just studies in their own right. They are fascinating in themselves, relevant to our immediate environment and very important economically when it comes to locating valuable deposits of oil and minerals. Astronomers can now use the studies of Earth, however, as a baseline for investigating the other planets. The disciplines of geophysics and geochemistry have enabled scientists to design instruments to study the rocks of Venus and Mars from unmanned spacecraft; while the meteorology of the Earth's atmosphere has helped to explain the convoluted patterns of clouds and whorls in Jupiter's deep atmospheric layers.

Geophysicists can study the Earth's deep interior only by indirect methods. Our globe is 12,756 kilometres (8000 miles) in diameter, and the deepest mines have penetrated only $3\frac{1}{4}$ kilometres (2 miles) into it. Fortunately, there is a way of 'X-raying' the Earth. Vibrational (seismic) waves from earthquakes spread out through the Earth's interior, at a speed which depends on the properties of the rock they are passing through. Some types of earthquake wave – the S-waves – will not pass through liquids at all. By analysing the earthquake records at seismometer stations around the world, scientists can deduce the properties of rocks deep inside the globe.

These seismic waves show that the Earth has three regions. There is a surface *crust*, of fairly lightweight, solid rocks. It is 35 kilometres (22 miles) thick under the continents, and only 5 kilometres (3 miles) under the ocean floors. Below lies the *mantle*, denser rocks that make up the bulk of the Earth. And right at the centre is the Earth's *core*, a region 7000 kilometres

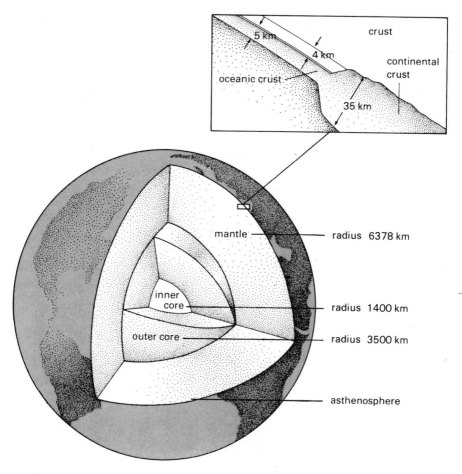

2 INTERIOR STRUCTURE OF THE EARTH *The interior of the Earth comprises several concentric layers. At the centre is a small core of solid metal, surrounded by a layer of liquid metal in the outer core. Above the core lie the highly-compressed rocks of the mantle, which constitute the bulk of our planet. The solid surface crust floats on the outermost layer of the mantle, the semi-liquid asthenosphere. The extra height of Earth's continents is matched by a greater depth of crust underneath, making the continental crust considerably thicker than the crust of the ocean floor.*

(4375 miles) in diameter. The outer part of the core must be liquid, because it blocks off S-waves travelling through the Earth, but the central part is probably solid. The core must be much denser than the mantle, to account for the Earth's total mass, and scientists believe it is made of metal. To judge by the composition of metal meteorites, which are fragments from the cores of disrupted minor planets, the Earth's core consists of an alloy of iron and nickel.

Moving streams of metal in the liquid core generate electric currents, and these in turn produce a magnetic field, in exactly the same way that electric currents circulating in the coils of an electromagnet produce its magnetism. It is as if our planet has a short bar magnet in its centre. The magnetic field extends out through the Earth's surface, to a distance of some 100,000 kilometres (62,500 miles). It acts as a shield against the wind of electrically-charged particles – electrons and protons – which continually stream outwards from the Sun. Most of these *solar wind* particles are deflected around the Earth, but some manage to penetrate it and become trapped, in radiation belts which surround the Earth above the equator. The most intense of these doughnut-shaped regions are the two Van Allen belts. They were an unexpected discovery by the first American satellite, Explorer 1, and lie 300 and 22,000 kilometres (188 and 13,750 miles) above Earth's surface.

Particles of the solar wind can penetrate right down to the Earth itself at the two magnetic poles, where the magnetic field is vertical. As the particles speed downwards, they strike the atoms of the upper atmosphere, causing them to glow in shifting patterns of greens and reds. These brilliant displays

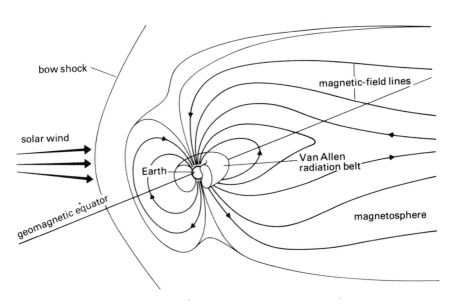

3 EARTH'S MAGNETIC FIELD REGION *The Earth's magnetic field stretches out from the magnetic poles like the field around a toy bar magnet. However, the Earth's field is enclosed by the solar wind of electrically-charged particles from the Sun; instead of reaching out indefinitely, the Earth's magnetism is limited to a region called the magnetosphere. The solar wind compresses the magnetosphere on the sunward side, while on the other side the magnetic field stretches out in a long magnetotail. Some solar wind particles penetrate into the magnetosphere and are funnelled down towards the magnetic poles. If they hit the atmosphere, these particles generate the light of an aurora; alternatively, they may be diverted into a reservoir of charged particles around the Earth's equator, called the Van Allen 'radiation belt'.*

of *aurorae* are often visible at night in polar regions. Above the north magnetic pole they are called the Aurora Borealis or Northern Lights; the corresponding displays at the south magnetic pole are the Aurora Australis, the Southern Lights.

Interestingly enough, the Northern and Southern Lights are exact mirror images of each other, right down to individual 'folds' in the great curtains of light and the constantly-changing shapes. The solar particles do not just plough into the air and lose all their energy at once; they actually oscillate back and forth along the lines of magnetic field high above the atmosphere, dipping into the air briefly at each end. Each individual electron is involved in painting two aurorae at once, separated by the middle of the Earth, and constantly whizzes from one to the other every second or less.

Earth's magnetism can be felt at its surface, by its effect on small magnets. Navigators traditionally rely on compass needles – small suspended magnets – to show the directions north and south. The magnetic poles however do not coincide precisely with the true north and south poles of the Earth, the points at the end of the imaginary axis passing through our rotating planet. The magnetic pole in the Arctic, for example, lies in northern Canada, some 1500 kilometres (940 miles) from the North Pole itself. And the magnetic poles are gradually moving, so causing compass bearings to alter slowly all over the Earth.

The changes are not really surprising, for Earth doesn't have a fixed magnet in its core, but an 'electromagnet' whose strength and orientation vary with the whims of the flows of molten metal within it. Analysis of ancient rocks even reveals that the Earth's magnetism has often 'flipped over' – so that a compass needle would swing around to point exactly the opposite way. These reversals are not regular occurrences, happening at random some three or four times every million years. When a reversal happens, the Earth's field gradually decreases to zero, then builds up again with field in the opposite direction. There is a period of a few thousand years when the field is so weak that it cannot shield us from the perpetual assault of the particles streaming from the Sun; and during such a reversal, Earth's atmosphere bears the full brunt of the solar wind.

When the Sun flares into high activity during a magnetic field reversal on Earth, the aurorae must stretch across the whole sky, from one pole across the equator to the other pole. And the influx of high energy radiation may have serious consequences for life. Even if the solar particles themselves are stopped in the atmosphere, their collisions generate more radiation and high speed particles which irradiate the ground below. Such radiation and particles can damage living cells. Most vulnerable are the reproductive cells, for even a small amount of damage here can produce serious changes in the next generation.

In the long run, such genetic changes – mutations – are beneficial to the progress of life on Earth. Animals and plants are always competing for food, for light, for space, and only the 'fittest' will survive to produce another generation. Most mutations will upset the delicate balance of life, and produce offspring which will die before reaching maturity – either as a direct result of the mutation, or because they are handicapped in the survival stakes. But very occasionally, a mutation will give the offspring some advantage over their parents and kin – a better turn of speed, a longer neck for grazing off trees – and the descendants of this mutant will carry this new characteristic through to later generations, and in time produce a new species. In fact it is only by random mutations that living things have been able to change from simple cells to the complexity of life-forms today.

Many factors can cause genetic damage to cells, and radiation from space is one of them. While we are protected by the invisible shield of Earth's magnetic field, we are relatively safe. But at times of magnetic reversal, radiation damage is a much more serious danger. The cumulative effect over hundreds of reversals has probably been an important contribution in evolution. Through its magnetic effects, the Earth's core has had an indirect, but abiding influence on the life forms which inhabit its surface.

Above the core is the great region of Earth's mantle, a rocky layer which makes up the bulk of our planet. Geologists believe it is made of a rock-type called peridotite, a silicate of iron and magnesium, with other metal atoms in smaller proportions. The temperature probably rises to several thousand degrees in the lower mantle, high enough to melt or even vaporize rock under normal conditions. But here conditions are far from normal: the weight of overlying rock layers compresses the mantle rocks at over a million times atmospheric pressure, and this pressure is sufficient to keep them almost solid. The mantle behaves like a solid as far as earthquake waves are concerned, but over longer periods of years or centuries its rocks can gradually flow like a very viscous liquid. Mantle rock is similar to pitch in this respect: a block of pitch will smash like a brittle solid when hit by a hammer, but left in a funnel it will very gradually flow out. Radioactive atoms – such as uranium – in the mantle rocks keep it hot, and because the Earth is coolest near the surface where the heat is lost to space, the mantle is rather like a pan of water on a low heat. The heated rocks at the bottom become less dense, and so they rise in slowly ascending 'plumes'. As the rocks reach the top of the mantle, just below the crust, they cool down, become less dense, and sink back towards the centre of the Earth.

The very uppermost layer of the mantle is the least solid. This region, lying just below the crust, is known as the *asthenosphere*, or more graphically as the 'slush layer'. The asthenosphere has been outstandingly important in Earth's long history, because its slipperiness has meant that the thin, solid crust on top can move about. The crust is in fact broken up into several large 'plates' which are in continual motion. Some of them include ocean floor, while other plates carry continents on their backs. The continual motion of Earth's crustal plates, and the interactions at their edges, have played the major role in shaping the Earth's surface – altering the distribution of land and sea, raising mountain ranges and stoking up the fires of volcanoes. This description of geology is called *plate tectonics* and is a relatively recent theory. It incorporates the older idea of 'continental drift', but expands on it – explaining *why* the continents should move about the globe.

According to the theory of plate tectonics, the surface of the Earth consists of half-a-dozen major plates, with several smaller plates tucked between them. Some of the plate names are self-explanatory: the Eurasian, the African, the Antarctic and the American. Each of these plates includes the adjacent sea floor, as well as the continents themselves, so that the American plate, for example, borders the Eurasian and African plates along the centre of the Atlantic Ocean. The Indian plate comprises the floor of the Indian Ocean, and two land masses – India at one end, and Australia at the other. The sixth of the major plates is the Pacific, which consists entirely of ocean floor.

As the rocks in the mantle gradually flow up, across and down again, they drag the surface plates about. Where two plates are separating, fresh molten rock oozes up from the slush layer below to fill the gap. The rock solidifies as a ridge at the plate junction; and as the rupture continues the new rock splits, half attaching to each plate edge, and yet more molten rock rises between. Thus plates can grow in size as new rocks are added at their margins. The best example of a zone where plates are separating on our planet is the mid-Atlantic ridge, a long submarine mountain chain which winds its way exactly halfway between the Americas on the west, and Europe and Africa to the east. The plates on opposite sides of the Atlantic are moving apart at a rate of two or three centimetres (about an inch) per year, and as they separate molten rock wells up at the mid-Atlantic ridge.

The South Atlantic ridge provides the most straightforward 'proof' of plate tectonic theory, because it displays several independent pieces of evidence which all tie neatly together. For a start, South America and Africa *look* as though they have broken apart: on a globe, we have all noticed how the 'bulge' of South America would fit into the 'hollow' of west Africa. If we could put these jigsaw pieces back together, we would find that the rocks of Africa fit exactly to similar outcrops in South America. And fossils show that the two continents used to share an almost identical range of living creatures and plants, including many organisms that could not have crossed a wide ocean.

The fossils suggest that Africa and South America parted company some 100 million years ago. This is backed up by direct dating of the ocean-floor rocks. Analysis of radioactive elements tells geologists how long ago the rocks solidified, and the whole of the Atlantic ocean floor turns out to be less than 100 million years old – very young in terms of the Earth's 4600 million year past. The most compelling evidence however lies in the magnetism of the ocean-floor rocks. As molten rock cools, it acquires a magnetism from the Earth's magnetic field. According to the theory of plate tectonics, the Atlantic ocean floor has been created gradually and continuously from the centre. As a result, the present mid-ocean ridge is the youngest part, forming even now, while rocks farther to either side are progressively older. The ocean floor immediately adjacent to the continents must date back to the time when Africa and South America were beginning to split apart. So the succession of rocks from the centre outwards should bear the imprint of the Earth's magnetism back in time, like a giant tape recorder.

Since rocks from the continents have shown that the Earth's magnetism suffers reversals from time to time, we should expect the ocean floor rocks to be magnetised in alternate 'stripes' of opposite magnetic direction, as we go from central ridge to either edge. If the ocean has been spreading at a constant rate, the width of the stripes should correspond with the duration of each magnetic reversal. And because the new rocks spread out on both sides of the ridge, each half of the ocean bed should be an exact mirror image of the other.

With hindsight, such ideas seem obvious. Twenty years ago, however, the theory of plate tectonics had yet to be formulated, and very few geologists believed in 'continental drift', even though evidence was piling up that the ocean floors were unexpectedly young. Unlike most changes in scientific thought, the revolution came quickly, and dramatically. In 1960, no respectable geologist would admit continental drift occurred; by 1970 only a few reactionaries still held a belief that Earth's surface is constant and immutable. What swung opinion so strongly was the magnetic pattern on the sea floor. About 1960, American scientist Harry Hess made the then-heretical proposal that ocean floors spread outwards from the mid-ocean ridges, but he lacked conclusive proof and few scientists took the idea seriously. One who did was English geologist Fred Vine, who first realised that the ocean floor should bear the imprint of magnetic reversals as tell-tale 'strips'. In 1963, Vine and his colleague Drummond Matthews published their first interpretation, of stripes in the Indian Ocean; and by 1967 they had conclusive proof from the Atlantic Ocean floor. Its rocks *do* bear magnetised stripes, and the stripes are mirror images on the two sides of the mid-ocean ridge. In addition, their widths match precisely the known durations of the magnetic reversals. With later results from other ocean floors, there was no room for doubt: the continents do move about the globe.

A significant problem cropped up immediately. If some of the Earth's plates are increasing in size, as rocks are added to them at the mid-ocean ridges, then other plates must be decreasing in area to compensate. Where do these rocks go? The answer is not too difficult to find, and it in fact solves a number of other geological problems – in particular why there is a 'ring of fire' around the Pacific Ocean, an almost continuous chain of volcanoes running from the Andes and Rockies through the Japanese Islands and down to New Zealand.

Where two adjoining plates are moving together, something must give. In fact, the only solution is for one to ride up over the other, and the lower plate is forced down into the mantle where it melts. Such regions are called *subduction zones*. South America for example is moving westwards, and is riding up over one of the smaller plates, the Nazca plate, which floors the Pacific at this point. As the rocks of the Nazca plate are melted beneath western South America, the lighter molten rocks force their way up through the overlying continent and burst out as volcanoes – the long towering chain of the Andes. The western edge of the Pacific plate is being forced under the Eurasian and Indian plates, giving rise to the volcanoes of Japan and the East Indies. The over-ridden plates do not give in without struggle, though, and their cracking up is felt above ground as the earthquakes which are all-too-familiar around the Pacific.

The rocks which make up continents have a lower density than ocean rocks, however, and this means that continents never suffer the indignity of subduction. When continental plate meets ocean floor plate, the continental plate always comes out on top – literally. If two continents meet, then the energy of collision is absorbed in a crumpling-up of the rocks themselves, into huge mountain ranges marking the collision line. The Himalayas, for example, are the crumpled debris of two continent margins, forced up into towering piles as the subcontinent of India has cannoned into the southern edge of the Eurasian plate.

The continuous recycling of ocean floor rocks, from solidification at mid-ocean ridges to re-absorption in subduction zones, means that nowhere on Earth are ocean floors more than 250 million years old – only 6 per cent of the Earth's age. The continents consist of older rocks, but they have also been renewed, in a more gradual way. Part of their matter consists of new rocks belched out by volcanoes – like those of the Andes – but the rocks have generally been rejuvenated by ordinary geological processes, particularly by the erosion of running water. Flowing streams and rivers, and the more stately glaciers, cut away the old rocks fragment by fragment, and deposit them in shallow seas as sedimentary rocks. The forces of plate tectonics eventually raise these rocks to become dry land once again – but of a younger age.

The processes of plate tectonics and erosion have between them wiped out the original surface of the Earth. The oldest known rocks, from Godthåb in Greenland, date back only 3800 million years, to a period when the Earth was already 800 million years old. The Earth's surface originally must have been heavily cratered, like those of our neighbouring worlds Mercury, Venus, Mars and the Moon. Our planet appears so different superficially not because it formed in a unique way, but because erosional and geological processes have renewed its surface, continually wearing away existing landscapes and creating new ones.

In time-lapse photography, covering a million years a frame, we would see Earth's surface as a continually changing pattern of continents and seas. The shapes on the familiar school-room globe comprise only a snapshot of our world at one moment in time: the familiar continents only fleeting features of our planet. Continental drift has been going on for at least the past 2700 million years – over half the Earth's lifespan – and it has had major effects on life on the Earth.

In the development of life by evolution, competition has been an essential ingredient. A 'fitter' mutation will only replace the existing variety if the two are competing for food or living space, and the resulting changes are usually gradual. But life forms evolved in different ways on the various continents of Earth, separated by wide ocean barriers, and when continents drifted together the effects on life have been dramatic. Around 150 million years ago, Earth's land-masses formed two huge super-continents. Laurasia comprised what is now North America, Europe and Asia, while Gondwanaland included South America, Africa and Australia. Mammals developed in both, but with different methods of nurturing their young. The Laurasian mammals were placental, retaining their foetuses within a womb, while the mammals of Gondwanaland were marsupials, nourishing the foetuses in an external pouch. Marsupials spread over South America, Africa and Australia. The present indigenous mammals of Australia – kangaroos, koala bears and so on – are marsupials. But after Gondwanaland broke up, Africa moved northwards to run into the southern part of Laurasia. The way was now free for marsupials and placental mammals to mix – with disastrous consequences for the African marsupials. In direct competition, they were less successful than the placentals, and rapidly died out over the whole African continent. The advantage for life in the long term was that the placentals suddenly had a new continent, with vast new reaches of living space and new habitats; their 'victory' over the marsupials gave the natural processes of evolution a chance to stretch out into new types of living creature.

With the rolling savannah grasslands before them, herbivores diversified into the many different varieties of antelope and gazelle, zebras, wildebeest and the giraffe, which escaped competition by developing a long neck to graze off trees. Carnivores developed to feed off the herbivores in various ways, from the tremendous speed of the cheetah to the brute force of the lion, and down to the pure scavenging of the hyena. Other animals took to the trees of the great tropical rain forests, to establish the lines of the monkey families – which eventually produced the big-brained, bipedal primate which calls itself *Homo Sapiens*.

Such continental rearrangements affect marine life too. Marine life generally belongs to the shallow seas off the coasts, not to the deep oceans between. When continents collide, these seas are forced up into mountain chains, and their living burden is destroyed; but the breaking up of super-continents creates whole new shallow seas at the continent margins, and new life-forms can proliferate into the new seas. In these ways, continental drift has undoubtedly speeded up the processes of evolution on our planet, to the extent that the intelligent species Man has appeared now – while the Sun is still shining. If the process of evolution had progressed at only half this rate, we should not have evolved for another 5000 million years – when the Earth will be about to shrivel up in the Sun's death-throes.

The Earth's atmosphere stands in exactly the opposite relationship to life. While living things are subject to the whims of shifting continents, it is living creatures that have created the Earth's unique atmosphere mix of nitrogen and oxygen. Venus and Mars have atmospheres made almost entirely of carbon dioxide gas, with only a couple of per cent of nitrogen and virtually no oxygen. A comparison with the gas giants, like Jupiter, is meaningless because their greater gravitation has held on to a large amount of the lightest gases, hydrogen and helium, but Saturn's moon Titan – the only satellite with an atmosphere – has an atmospheric blanket of almost pure nitrogen. Titan's atmosphere is essentially the same as that of Mars and Venus, with the difference that the expected huge amounts of carbon dioxide are frozen solid at this distance from the Sun – indeed, this frozen gas must make up much of the solid body of Titan itself.

Earth's atmosphere thus holds two puzzles. One is the lack of much carbon dioxide – it constitutes only 0.03 per cent of our air – and the other the presence of huge amounts of oxygen, which is such a reactive gas that we should not expect it to remain uncombined in an atmosphere for billions of years. The answers once again do not depend on the Earth having been born different from our neighbouring worlds. Our planet has altered its original atmosphere.

Earth's first mantle of gases could have been formed either from the exhalation of volcanoes, or from the evaporated bodies of comets and other icy debris which fell on the Earth in its early days. In either case, the major constituents must have been molecules of water vapour and carbon dioxide (a *molecule* is a chemical combination of atoms), with smaller amounts of nitrogen, and of noxious gases like carbon monoxide, hydrogen sulphide, ammonia and hydrogen cyanide. Earth's atmosphere began to differ from those of its neighbours even at this stage. Our planet lies at just the right distance from the Sun for water vapour to condense to ordinary liquid water – and this it did. For countless millions of years, thick overcast skies must have drenched the Earth with continuous rainfall, until the spaces between the shifting continental blocks were filled with liquid oceans to a depth of 4 kilometres (2.5 miles). And liquid water can dissolve carbon dioxide. As the oceans filled up, they dissolved the vast majority of the carbon dioxide molecules in the atmosphere. Some remained in dilute solution in the seas; but much of the dissolved carbon dioxide reacted chemically with Earth's original rocks, and became locked up in carbonate rocks. The atmosphere became much less dense, till it was as much nitrogen as carbon dioxide.

The condensation of water has made our world unique. On Venus, lying close to the Sun, the temperature was always above the boiling point of water, so the vapour stayed in the atmosphere and has eventually been lost to space. This planet is still enveloped in a thick mantle of carbon dioxide, with clouds of poisonous, sulphurous compounds. On chilly Mars, the water vapour froze out as solid ice, once again to leave a carbon dioxide atmosphere.

The evolution of our atmosphere from this point has been in the hands of living things. Plants derive their energy from sunlight, and their 'food' is carbon dioxide from the atmosphere. In the process of photosynthesis, the plants absorb carbon dioxide, use its carbon atoms to grow and reproduce, and reject oxygen atoms from the carbon dioxide molecule as waste. Over billions of years, early plants on Earth gradually transformed the entire atmosphere of our planet from a mixture of the inert gas nitrogen and carbon dioxide to its present composition of nitrogen and oxygen.

We now have a state of balance, in fact, because animal life breathes in oxygen, and exhales carbon dioxide – and uses as 'fuel' the carbon from the plants, or other animal bodies, that we eat. Animal life is exactly like a controlled fire. While a careless match will set a wheatfield blazing in wasteful flames and smoke, our bodies burn wheat ears in basically the same way when we eat a slice of bread. In either case, the wheat is combined with oxygen to generate energy, while carbon dioxide is emitted.

Plant life has however established that oxygen is the more abundant gas in our atmosphere. Not only does this allow animals to breathe, but it has been vital for the development of life on land. The Sun bombards the Earth continuously with dangerous ultraviolet radiation, but this is blocked off by a layer of oxygen some 50 kilometres (30 miles) up, called the *ozone layer* (because the oxygen here is bound up in the molecular form *ozone* which contains three oxygen atoms). The layer is a necessary armour for all but the ocean-dwelling creatures (see page 6). But the ozone layer is vulnerable. It is thin, and it lies high in the atmosphere. Influences from space beyond may be able to destroy it, and so unleash the Sun's ultraviolet rays on to the Earth beneath. As we shall see in later chapters, some scientists believe that such events have happened in Earth's past – and will happen again. And there is a danger that we ourselves may damage Earth's thin protective layer by the action of artificially produced chemicals. In particular, the fluorochlorocarbons that power aerosol sprays can rise up through the atmosphere and react chemically with ozone to reduce the protective effect of the ozone layer.

There is another possible danger inherent in Man's exploitation of Earth.

In breaking up the original carbon dioxide atmosphere, plants have locked up an enormous amount of carbon. Some of this is contained in the plants which are alive today, but much more is buried, as fossils of the plant life of ages ago. The richest repositories of fossilised carbon are the coal beds and oil deposits. As we burn coal and oil for energy today, we are squandering millions of years-worth of the plants' efforts to create an oxygen atmosphere. In the burning of fossil fuels, we are gradually decreasing the amount of oxygen in the atmosphere, and replacing it with carbon dioxide. Ultimately, if we were to burn the whole of present plant life and all the fossil fuels, we would convert all the oxygen back to carbon dioxide.

There is no real danger of that. But scientists have found that the carbon dioxide is increasing at a measurable rate – by about ten per cent over the last century. From one point of view, this does not seem significant, because oxygen still outnumbers carbon dioxide 700 to one – but the change is more serious than these figures suggest. Carbon dioxide is very efficient at trapping the Sun's heat (like the glass of a greenhouse), and even a small increase can alter the world's temperature significantly. And even a slight heating of the Earth, by less than a degree, will change the climate of vast areas – extending the deserts over thousands of square kilometres of now-fertile land, bringing starvation to millions, and possibly the threat of war to already unstable countries.

To offset this artificial warming of the Earth, however, is the fact that world temperatures are gradually falling as we head towards another glaciation (period when ice sheets will cover much of the temperate regions). For the last two million years the Earth has been in the grip of the Great Ice Age, but the ice cover has not been constant. Its extent has varied – from just the polar regions, as at present, to ice sheets extending down as far as London and New York. We are living now in one of those relatively warm 'interglacial periods' when the ice has retreated temporarily. The last major glaciation ended only 10,000 years ago; and it is certain that the ice sheets will advance once again within the next 10,000 years – unless Man intervenes, either unintentionally by increasing the carbon dioxide in the atmosphere and thus warming the planet, or by some sophisticated means that another hundred centuries of technology will bring to us.

The prediction of a future glaciation is not just crystal-ball gazing, or the simple extrapolation of past glaciations and interglacials. In the past decade, scientists have come to understand *why* the ice sheets ebb and flow. Their movement is caused by changes in the motions of our planet itself. Astronomers have suggested the possibility for a century, but the changes are so small that few climatologists believed that they could lead to wholesale changes in our planet's temperature. The Yugoslav scientist Milutin Milankovitch put forward the theory in its present form in the 1930s, and in the past decade geologists have established that his calculations of temperature changes agree precisely with evidence obtained from deep-sea sediments, which preserve a record of ocean temperatures.

Milankovitch did not propose that the total amount of sunlight reaching the Earth each year was any different during glaciations and interglacials. His theory depends on the severity of the seasons, which is related in turn to the tilt of the Earth's axis, and to the shape of the Earth's orbit. The Earth's tilt causes the seasons, because the hemisphere tilted towards the Sun receives sunlight more directly and so is warmer – it experiences summer. Simultaneously, the other hemisphere is tilted away, and has winter conditions. Every six months, the Earth is the other side of the Sun, and the situations are exactly reversed.

The Earth's axis can change its tilt, nodding up and down by a few degrees over a period of 40,000 years. When it is least tilted, the seasons are moderate, with cool summers and mild winters. The *direction* of tilt matters, too, because the Earth's orbit is not circular. If the Earth's axis is orientated so that the months of winter and summer (in either hemisphere) occur when Earth is at its closest approach to the Sun and its furthest point, then the difference in distance from the Sun will exacerbate the normal seasonal changes and lead to large contrasts between summer and winter temperatures. The axis direction swings around with a period of 26,000 years. Finally, the shape of the Earth's orbit is important, because the last effect will be most marked when the orbit is most elongated, and least important when the orbit is nearest to a circular shape. Although the Earth's orbit should stay constant in shape, the gravitational pulls of the other planets do cause it to gradually alter, from a relatively fat oval to a thinner shape and back, over a period of 100,000 years.

Milankovitch went on to explain how the seasonal contrasts can alter the Earth's overall climate. The critical period is during a time of mild winters and cool summers. Although only a little snow will fall at high latitudes at such times, the summer will not be warm enough to melt it away. As a result, the snow-cover of the Earth does not just come and go with the seasons: it builds up inexorably, increasing in size every winter.

As the bright, snow-covered areas of Earth increase, their white surfaces reflect more of the Sun's light and heat into space. The heat available to warm our planet thus decreases; its average temperature falls until the summers become not just cool, but cold – while the winters become frigid. More snow falls, and ice begins to extend down to the lower latitudes – a glaciation has begun.

When the Earth's orientation has changed, so that opposite conditions prevail, the relatively warm summers melt more ice than the severe winters can produce. Our planet begins to warm up, and an interglacial has begun. We are lucky enough to be living in such conditions now.

By combining the periods involved in the three postulated causes, it is

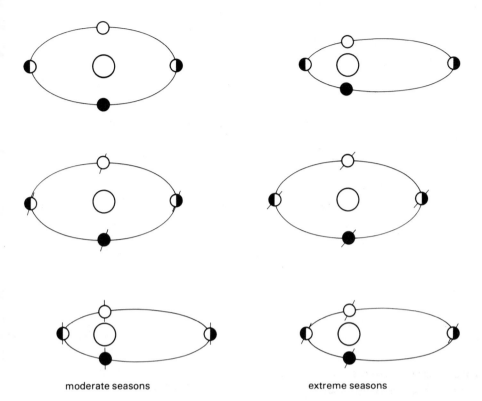

moderate seasons   extreme seasons

4 EXTREME SEASONS RELATED TO THE EARTH'S TILT AND AXIS *According to the Milankovitch theory, the recurrence of glaciations in an Ice Age is caused by changes in the Earth's orbit and in the tilt of its axis of rotation. Over a period of some 100,000 years, the Earth's orbit changes from near-circular to appreciably elliptical and back again (top). The Earth's axis 'nods' up and down with a period of 40,000 years (centre); and it is also swinging around in space, completing one turn in 26,000 years (bottom). The extremes of the three cycles, shown in the right-hand column, each lead to a very marked contrast between the seasons, with very cold winters and very hot summers. The other extremes (left-hand column) produce 'moderate' seasons with little contrast between summer and winter. During a period of moderate seasons, winter snow does not melt in the cool summer months, and it can accumulate over the years to produce a major glaciation.*

possible to calculate how the contrast between summer and winter temperatures has changed, and hence how the Earth's overall temperature should have altered. Measurements from ocean sediments has confirmed that the glaciations do indeed follow the Milankovitch pattern. And the corollary is that we can predict when the next glaciation will occur.

The vindication of the Milankovitch theory emphasises just how Earth is influenced by its neighbours in space. Our climate does not just depend on spaceship Earth itself, or the terrestrial atmosphere and oceans. Nor is the influence of the Sun the whole story. Our 'spaceship's' orientation and tilt change because of gravitational pulls on its bulging equator, and the principal culprit is the Moon. The shape of our planet's orbit is the other factor, and the alterations in shape are due primarily to the attraction of the nearest planet, Venus, and of Jupiter – whose great mass makes up for its greater distance.

There are also longer-term cosmic effects on Earth. The Great Ice Age which has been with us two million years is a relatively cool period of Earth's history, during which the Milankovitch process can work. Before this, the Earth was generally warmer, and no glaciations advanced over the temperate lands. Geologists have however traced earlier periods of glaciations, separated by intervals of a hundred or two hundred million years. Scientists still dispute why the Earth has suffered such Great Ice Ages, spread at enormous intervals throughout its past. Continental drift must be one factor, because ice and snow will accumulate much more readily when there is land near the poles. But the basic causes may well lie beyond the Earth, cosmic influences perhaps related to our membership of the huge galaxy we call the Milky Way. We shall take up some of these suggestions later (see Chapter 11).

Despite the Great Ice Ages, Earth's average temperature has never strayed more than a few degrees from its present, fairly pleasant temperature. This is rather strange, because astronomical theory tells us that the Sun has warmed up by almost fifty per cent during this time. Earth's temperature has remained constant only because the plants have been removing carbon dioxide from our atmosphere, and the decreased trapping of heat has exactly balanced the increasing heat from the Sun.

This is a remarkable coincidence – or is it? British chemist James Lovelock believes it is more than mere good luck. Earth's living things are all linked together by the common medium of our atmosphere – animals inhaling oxygen and breathing out carbon dioxide, which plants convert back to oxygen – and Lovelock regards all living things, the atmosphere and the oceans as essentially one living entity. He names this planet-wide entity *Gaia* – after the Greek earth goddess.

Gaia is more than just the sum of her parts. Every part of her is dependent on every other, and often on components of the total being that are totally inconspicuous. For example, plants and animals require compounds containing nitrogen atoms, and although nitrogen molecules constitute most of the air, there are only a few organisms which can convert it to a consumable form. On these few microscopic bacteria the rest of life on this planet depends. Again, the decay of dead plants and animals produces methane – marsh gas. If this built up in our atmosphere, then eventually it would form an explosive mixture – and an odd flash of lightning would blow up the entire atmosphere! Some of the least-known bacteria, which reside in muddy river estuaries, are equipped to 'eat' methane, and perform the important, but thankless task of keeping the atmosphere safe. There must be a multitude of other specialised bacteria, many not yet recognised, which keep our planet habitable for higher organisms like us – as essential a part of Gaia as we ourselves.

Lovelock thinks that Gaia may have continually altered the relative levels of carbon dioxide and oxygen to keep her temperature constant and compensate for the Sun's increasing luminosity. Although Gaia is not 'intelligent' in our normal sense of the word, she may have a 'sense of survival' which has seen her through for $3\frac{1}{2}$ billion years. The planet Earth is then not 'ours': we are a part of a larger being, and we tamper with other life and our atmosphere at our own peril. Burning fossil fuels and cutting down the tropical forests may cause Gaia to make an adjustment to compensate. Just as a man may amputate a gangrenous leg to save the rest of his body, so Gaia may regard mankind as an interesting experiment in evolution – but totally expendable if man should threaten the well-being of Gaia as a whole.

— 3 —

The Moon

'That's one small step for a man, one giant leap for mankind.' On 21 July 1969, these words crackled back to Earth over a third of a million kilometres (quarter of a million miles) of space, as Neil Armstrong became the first human to set foot on another world.

On the cosmic scale, the Apollo flights to the Moon were only a walk around our back garden. The Moon is the other half of the Earth-Moon double planet system, bound in permanent proximity to our world. Even the nearest of the other planets, Venus, never approaches closer to us than a hundred times the Moon's distance. Remote Pluto, at the edge of the Sun's family, is 16,000 Moon distances away – while the nearest star is a hundred million times more remote than the Moon.

Even so, we should not underestimate the importance of the manned Moon expeditions. They have proved that Man *can* leave the planet of his birth, and travel to other worlds. If the United States had continued to support manned space flight at the intensity of the Moon expeditions, we would now be seeing preparations for the first manned flight to Mars.

The American commitment to send a manned expedition to the Moon was politically inspired. Stung by the Russian's success in launching the first Earth satellite, and the first astronaut, Yuri Gagarin, President Kennedy committed his nation to reaching the Moon first. The Apollo 11 landing assuaged American pride; and after its success, interest in space waned rapidly – particularly when this expensive pastime was hit by the recession of the 1970s.

But the six successful Apollo landings were not just an expensive political gambit. The astronauts brought back valuable scientific information about the Moon, including a third of a tonne of lunar rocks, and showed that a manned expedition was far more versatile than unmanned landers of the kind that the Russians had used. Australian Moon expert Stuart Ross Taylor says: 'Could our insights into the composition, age and evolution of the Moon have been obtained from unmanned missions? The answer is no.' The last Apollo flight, for example, included a geologist, Harrison 'Jack' Schmitt, who could inspect the surface at close quarters, and choose the most interesting rocks to return to Earth for detailed analysis. The Apollo astronauts also left instruments on the lunar surface, including seismo-meters to detect any ground-shaking due to 'moonquakes' and instruments to measure the amount of heat flowing upwards from its centre.

As expected, the Moon turned out to be a dead world – but an invaluable museum of the early history of planets. Because the Moon is too small to hold on to an atmosphere, its surface is exposed to the rawness of space. It is shotblasted by tiny meteors and solar wind, and alternately baked to 110 °C (230 °F) – hotter than boiling water – during the lunar day and chilled to −170 °C (−274 °F) during the long lunar night. But the rugged surface has not been worn down by erosion or disturbed by the geological forces that have renewed most of the Earth's continents and ocean beds. Most of the Moon is solid rock. Its surface is not broken up into shifting plates like the drifting continents of the Earth which constantly reprocess the rocks of the Earth's surface. And because the Moon has no atmosphere, it has no water to erode its mountains and build up sedimentary rocks. The surface of our companion world has remained unchanged for thousands of millions of years.

When we look at the familiar face of the 'Man in the Moon', we can see at a glance that the Moon is a patchy world, with two different kinds of surface: superimposed on a brighter background are darker regions making up his 'eyes', 'nose' and 'mouth'. Even a small telescope shows that the brighter regions of the Moon are mountainous *highlands*, while the darker parts consist of low-lying flat plains. Because the early astronomers thought these plains were seas, they are known by the Latin term *maria* (singular *mare*) to this day. The smooth mare surfaces are broken by occasional low meandering ridges and narrow twisting gorges (*rilles*), while several of the more-circular maria are surrounded by high mountain walls – some as high as Mount Everest, on a world a quarter the Earth's size! The plains bear fanciful names from an earlier age of astronomy: Mare Foecunditatis, the Sea of Plenty; Oceanus Procellarum, the Ocean of Storms; and Sinus Iridum, the Bay of Rainbows.

But the most striking feature of the Moon, seen through a telescope or from a spacecraft window, is its craters – circular walled enclosures ranging in size from huge walled plains 250 kilometres (156 miles) across down to small saucer-shaped depressions only a metre (3 feet) in size. In the highland regions, the craters stand packed together, jumbled and overlapping so that there is hardly a smooth patch of ground, while even the smooth maria are finely peppered with medium and small craters. The larger craters have been named after famous astronomers, philosophers, and scientists, including Copernicus and Plato. Because the craters were named three centuries ago, the largest craters often bear the names of astronomers now otherwise forgotten, like Clavius and Grimaldi, while later, more eminent scientists have had to make do with smaller and more obscure craters. The crater Einstein, for example, is right at one edge of the

Moon as seen from Earth, so it is difficult to pick out even under the best conditions.

Astronomers have now been able to piece together the history of the Moon from the discoveries made by the Apollo missions, and explain the origin of the highlands, the maria and the craters. Before men went to the Moon, scientists were also confident that a detailed study of lunar rocks would answer the most fundamental question of all: where was the Moon born? Oddly enough, that question remains unanswered. Scientists have been able to rule out one possibility, the idea that the Earth and Moon were once one single planet which was spinning so rapidly that it split into two. This theory would mean that the Moon's rocks should be very similar to those of the Earth's surface layers, but the samples brought back by the Apollo astronauts have turned out to be very different. The Moon contains a higher proportion of the elements titanium, uranium, calcium and iron in its rocks, and a much lower proportion of many other elements, including gold, silver, zinc, bismuth and potassium.

Astronomers are still hotly debating the other two theories of the Moon's birth. It could have formed as a separate planet, orbiting the Sun in its own right, but following an elongated path which crossed the Earth's orbit. On one fateful occasion, it passed too close to our planet, and the Earth captured it, pulling the Moon from its Sun-centred orbit to become a satellite of the Earth – or, more accurately, the Earth and Moon captured one another, to become a double planet. This theory naturally explains the Moon's different composition, for astronomers expect that planets formed at different distances from the Sun should not contain exactly the same mix of the elements (see Chapter 6). The capture theory also provides a neat answer to the puzzle that the Earth is the only planet of the four inner planets to have a large moon: it's pretty unlikely that two planets should capture one another anyway, and virtually impossible that it should happen twice in the same planetary system.

But some astronomers think that even the capture of the Moon by the Earth is 'horrendously improbable', to use the words of planetologist W.M. Kaula. The alternative theory says that the Earth and Moon formed together in space. The early Earth was surrounded by a ring of rocky fragments – rather like Saturn's rings today – and these rocks came together to make up the Moon.

Despite this controversy over the Moon's birth, astronomers accept that it formed at the same time as the other planets, 4600 million years ago, and they have a fairly clear picture of what has happened since. Loose chunks of rock pounded into the young Moon and built it up to its present size, and their impacts melted the surface layers down to a depth of several hundred kilometres. Lighter rocks floated to the surface of this incandescent ocean, and solidified there as a thick layer of scum. This ancient crust forms the highlands we see today, and it lies beneath the mare plains too. The hail of rocks from space still continued, but the bombardment gradually slackened off. Instead of melting the Moon's surface, the infalling meteorites simply blew out craters when they landed. By 3900 million years ago the bombardment had virtually ceased.

The molten ocean of rock had long since solidified through, but now a new source of heat was melting small reservoirs of rock below the Moon's surface. This was radioactive heat from elements such as uranium – the source of heat energy in man's nuclear reactors. The molten rock made its way up through cracks in the Moon's shattered crust, and spread out as lava flows over the low-lying regions of the lunar surface. Here they solidified to form the dark mare plains. These lava flows covered the old cratered crust below, creating a new smooth surface. Lava naturally welled up into the largest of the old craters, to make huge circular maria, like the Mare Imbrium (Sea of Showers) which is 1200 kilometres (750 miles) in diameter. Astronomers now recognise that the mountains and plains of the Moon are not in the slightest like the Earth's mountain and plain regions, formed by the geological processes of continental drift and erosion. The lunar mountains are simply the rims of enormous ancient craters, and they circle plains which are solidified lakes of lava that have flowed into the craters' deeply-dug interiors.

Very little has happened on the Moon since the last lava flows welled up over 3000 million years ago. Their deep reservoirs have now solidified, and the only parts of the Moon now molten are right down near the core. The Apollo seismometers have detected moonquakes in these deep regions, but they are a pale shadow of our earthquakes. British geophysicist G.H.A. Cole estimates 'the energy given off by moonquakes in one year is rather less than that involved in a good-sized fireworks display'. The surface has hardly changed at all, apart from the occasional new crater blasted out by the occasional large meteorite.

The Moon is an invaluable fossil of the early solar system. Other planets have altered themselves in one way or another: by erosion from winds or water, by volcanic activity or continental drift. But virtually everything interesting that's ever happened to the Moon was a direct result of its early environment. Understanding the simple, preserved story of our satellite world is a first step towards a full understanding of the other planets – including the Earth.

Although the Moon is only a very small body in the astronomical league – only one-eightieth as massive as the Earth – it is the only body apart from the Sun to affect our lives directly. We've all welcomed its light on a dark night when there's no other light-source about, and it was that much more important before streetlights became common. As the Moon follows its 27.3 day orbit around the Earth, the Sun illuminates the Moon at different angles, and as we see different amounts of the lit-up half, the shape – or phase – of the Moon seems to change. When the Moon lies on the same side of the Earth as the Sun, the unilluminated part is turned towards us and we cannot see the Moon at all. This is the time of New Moon.

As the Moon pursues its orbit, we see first a narrow crescent in the evening skies. Sometimes sunlight reflected off the Earth (Earthshine) can light up the dark part of the Moon's globe, and we see a complete, dimly-shining Moon edged by the brilliant sunlit crescent – the 'Old Moon in the New Moon's arms'. On succeeding nights we see more of the Moon illuminated, until Full Moon, when our satellite is opposite to the Sun and presents its lit-up face fully towards us. Then it shrinks back to a crescent, in the morning skies, before reaching New Moon again.

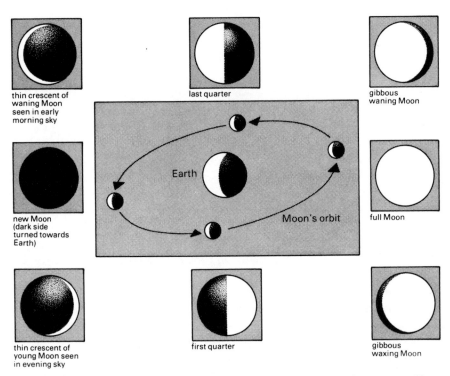

thin crescent of
waning Moon
seen in early
morning sky

last quarter

gibbous
waning Moon

new Moon
(dark side
turned towards
Earth)

Earth

Moon's orbit

full Moon

thin crescent of
young Moon seen
in evening sky

first quarter

gibbous
waxing Moon

5 THE PHASES OF THE MOON *The Moon's ever-changing phases arise from its monthly orbit about the Earth. We see the Moon by the sunlight it reflects; and as the angle between Sun, Earth and Moon constantly changes, we see increasing, and then decreasing, portions of the illuminated half of the Moon as it follows a complete orbit. The 'New Moon', directly between the Earth and the Sun, is dark because all the sunlight falls on the side turned away from the Earth. As the Moon moves round, the Sun illuminates a crescent-shaped part of the Moon facing us; this grows until the Moon is exactly half-lit as seen from Earth, the phase called 'First Quarter' because the Moon is then a quarter the way round its orbit. As it continues to move into more direct sunlight, the phase becomes increasingly bulging or 'gibbous'. Full Moon is reached when the Sun's rays fall directly on the face turned towards the Earth. After this, the phases run through in reverse order until the Moon is 'New' again.*

Although the Moon takes 27.3 days to circle the Earth – in other words, to return to the same point relative to the background of stars – the interval from New Moon to New Moon is slightly longer. The Earth orbits about the Sun in the course of a year, and as a result the Sun seems to move against the backdrop of stars. So after the Moon has completed one orbit, the Sun has apparently moved on a little, and the Moon has to travel that little bit further before it's in line with the Sun and is 'New' once more. The interval from New Moon to New Moon is 29.5 days – one lunar *month*, a word which was originally a 'moonth'. The lunar month is a very convenient unit of reckoning time, and some calendars – for example the Jewish calendar – are based as much on the succession of lunar phases as they are on the year.

The Moon has had a strong religious connection in many cultures, not only for its night-time light and its use as a time-keeper, but also because it causes the most awe-inspiring of all natural sights: an eclipse. At a normal New Moon, the Moon passes above or below the Sun in the sky: but sometimes it crosses right in front. By pure coincidence the Moon and Sun appear exactly the same size in the sky. When the Moon moves in front of the Sun, it completely blocks off its glowing disc, and darkness suddenly

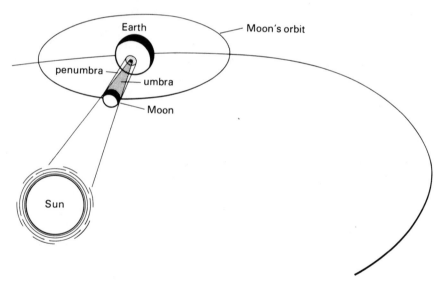

6 TOTAL ECLIPSE OF THE SUN *A total eclipse of the Sun takes place when the Moon comes directly between the Sun and the Earth and blots out the Sun's bright disc. However, eclipses are not as common as might be expected. Because the Moon's orbit is inclined to the Earth–Sun line, eclipses do not take place at every New Moon. And because the apparent sizes of the Sun and Moon are so similar, the alignment is precise only over a very small part of the Earth's surface (inside the Moon's dark shadow or 'umbra'). From a position in the outer 'penumbra' shadow, one sees only a partial eclipse of the Sun.*

falls in the middle of the day. Up in the sky, the outer atmosphere of the Sun becomes visible around the jet-black silhouette of the Moon. The sight of a total eclipse is strange and stirring even today when we understand its cause: in less-advanced cultures it naturally caused intense fear. One of the first duties of astronomers was to predict eclipses, and any astronomer who failed was likely to suffer the fate of the Chinese royal astronomers Hsi and Ho. The Emperor Chung K'ang beheaded this pair when they failed to predict an eclipse in 2137 BC. Some astronomers and archaeologists have proposed that many of the stone circles in Britain and north-west Europe were even earlier eclipse predictors, dating back to 2900 BC in the case of Stonehenge. The men who erected these stones may have used them to indicate particular positions on the horizon where the Moon would rise and set, and thus to help in calculating when eclipses would occur. The problem for us lies in working backwards to calculate which were the important directions at any megalithic site, and of course it is quite possible that a pair of stones might happen to lie along one of the directions purely by chance. In the 1970s, many scientists took the idea of 'megalithic astronomy' very seriously, and believed that our illiterate ancestors could make the very precise measurements needed to predict eclipses. But the more that the crucial stone alignments have been investigated, the less convincing they have become. There are undoubtedly a few real alignments, like the famous view of the Sun rising over the Hele Stone at Stonehenge on midsummer's day. But most of these are probably only rough markers, not the precision alignments needed to make eclipse predictions. They do show, however, that our ancestors had an active interest in the sky. British archaeoastronomer Clive Ruggles sums up current thinking when he says 'prehistoric astronomy may not be as sensational as has often been suggested, but it is no less intriguing for that.'

The Moon has a direct effect on those of us whose living involves the sea, for it is the Moon's gravitation which causes the tides. Strictly speaking, the Moon does not orbit the Earth, but both Earth and Moon are circling their common centre of gravity – their balance point. As far as the Earth is concerned its centre is moving around the centre of gravity at just the right speed for its centrifugal force to exactly balance the pull of the Moon's gravity. But the oceans on the side facing the Moon are 6400 kilometres (4000 miles) closer to the Moon, and so they feel its gravity more strongly – the Moon pulls them slightly upwards in a 'bulge'. The oceans directly opposite the Moon are less affected by the Moon's gravity, and here the centrifugal force is stronger. It pulls the waters in the opposite direction to the Moon's position, which is in fact away from the centre of the Earth. Hence centrifugal force raises a second tidal bulge on the opposite side of the Earth.

This tug-of-war between the Moon's gravity and the centrifugal force of the Earth moving around the Earth-Moon balance point thus makes the oceans take up a slight rugby-ball shape, with its length pointing towards the Moon. As the Earth spins round in twenty-four hours, the positions of the two tidal bulges seem to move around the globe, and each seaport experiences two 'high waters', and two 'low waters'.

In addition to their day-to-day importance to seafarers, the ocean tides are having a long-term effect on the Earth. They are gradually slowing down our planet's rotation. We can imagine the two tidal bulges as brake-shoes, held in line with the Moon, while the Earth spins between

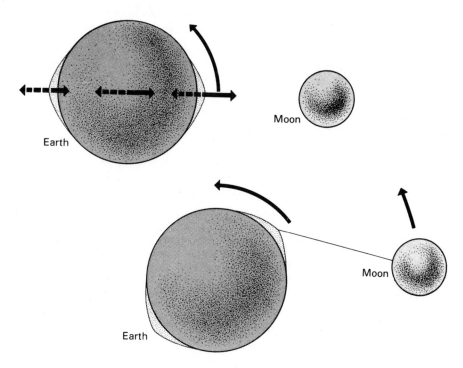

7 TIDES AND THEIR EFFECT ON THE MOON'S ORBIT *The moon's gravitational pull raises tides in the Earth's oceans, and these in turn affect both the Earth's rotation and the Moon's orbit. The Earth is in reality orbiting around the common balance point of the Earth–Moon system; as a result, every point on Earth feels the same centrifugal force (to the left in top diagram). At the Earth's centre, the force is exactly balanced by the Moon's gravitational pull (to the right). The Moon's pull changes with distance, though: on one side of the Earth (right-hand side above), it is stronger than the centrifugal force and pulls the oceans up into a high tide; on the opposite side, it is weaker and cannot prevent the centrifugal force from pulling the oceans into a second tidal bulge opposite the first. There is some friction between the water in the two tidal bulges and the solid Earth, which is rotating underneath them. The bulges act like giant brake shoes on the Earth, gradually slowing down its rotation and increasing the length of the day. Conversely, the friction pulls the tidal bulges ahead of the Earth–Moon line (below); the gravity of the bulges tends to pull the Moon forward, and as a result, the Moon's orbit is continuously increasing in size.*

them. There is slight friction between the tidal bulges and the Earth itself at the sea floors – very slight, but enough to affect the Earth's rotation to an extent that we can measure with modern atomic clocks. The length of the day is increasing at a rate of about one second in 50,000 years. This seems a small amount, but it accumulates inexorably over a planet's long history. We can calculate that during Devonian times, some 400 million years ago when life first evolved from the seas on to land, the day was only 22 hours long, and there were 400 days in a year (which was the same duration then as now).

The Earth raises slight tides in the solid body of the Moon, too. These tides have slowed down the Moon's rotation dramatically, so that whatever its original rotation rate may have been, it has now slowed down so much that one face of it is permanently turned towards the Earth. From here, we can only ever see one half of the Moon – the side with the 'face' on. The far side was completely unknown until spaceprobes and astronauts photographed it: intriguingly enough, it has turned out to be made entirely of bright highlands, with no big dark mare plains.

There's a final twist to the saga of the Earth–Moon tides. The Moon raises tides in the Earth's oceans, and these in their turn act on the Moon to increase the size of its orbit. This effect is easiest to understand by once again thinking of the tidal bulges acting as a pair of brake shoes on the rotating Earth. The frictional force between water and sea bed slows down the Earth, but it also means that the spinning Earth pulls the tidal bulges slightly ahead of the Earth–Moon line. If you looked at the Earth from the Moon, the high tide wouldn't lie directly below you, but displaced in the direction of the Earth's rotation – slightly to the east. The water in the oceans has a small gravitational pull on the Moon, and this effect means the Moon is continuously pulled very slightly eastwards as it follows its orbit. The result is that the Moon's orbit is very gradually increasing in size: the Moon is moving away from the Earth at a rate of 4 centimetres (1.6 inches) every year.

It's very difficult to measure directly such a small change in the Moon's distance, currently about 384,400 kilometres (240,000 miles). One indirect method is to look very precisely at the Moon's position in the sky: as its orbit gets bigger, the Moon should move slightly more slowly across the sky, and it should be always falling behind its predicted position. The most accurate way to keep track of the Moon's apparent laziness is also extremely simple. You simply record the precise time that the Moon moves in front of a bright star and hides it. The Moon *occults* several dozen bright stars in this way every year, and the only equipment you need to make a useful measurement is a small telescope – or even good binoculars – and a stopwatch. Thousands of amateur astronomers send in timings of lunar occultations every year, and the precise calculation of the Moon's orbital changes depend as much on these dedicated, unpaid amateurs as they do on the work by professional observatories. Leslie Morrison of the Royal Greenwich Observatory has recently made one of the most precise calculations, based on 62,000 observations of stars occulted by the Moon. The observations had been sent in over the previous 30 years by 2730 different astronomers – almost all of them amateurs.

Astronomers can now also measure the distance to the Moon very accurately by reflecting laser beams off reflectors left on the Moon's surface by the Apollo astronauts. The Moon is so far away that it takes $2\frac{1}{2}$ seconds for a flash of light to travel there and back: by measuring this time very accurately, and multiplying by the speed of light, scientists can calculate the distance to the Moon with an uncertainty of only 10 centimetres (4 inches) or so. The lunar laser-ranging scientists have indeed found that the Moon has moved away from us by about half a metre ($1\frac{1}{2}$ feet) since the Apollo moon landings of 1969 to 1972.

As the Moon's orbit gradually grows bigger, the length of the lunar month will gradually increase from its present duration of 27.3 days. But the day-length is increasing too, and at a much faster rate. Eventually, some thousands of millions of years into the future, the 'day' and the 'month' will be equal in length. The day will be fifty-five of our present days in length; the month will have the same duration, almost exactly twice as long as the month we have now. Because the Moon will move once round its orbit in the same time it will take the Earth to spin once around on its axis, our satellite will be moving in such a way that it always keeps station above the same spot on Earth. Descendants of ours who live here would see the small, dim disc of the distant Moon forever overhead, while those on the far side of the globe would never see the Moon at all.

In their distant waltz, the Earth and Moon will keep the same faces permanently turned towards one another, as the Moon keeps the same side facing Earth even now. The regions facing the Moon, and those directly opposite, will have permanent high tides; in other seaports on our planet the tide will always be out. The brake-shoes will be turning exactly with planet Earth, and tidal effects can no longer alter either the Earth's rotation or the Moon's orbit. Our 'double planet' system will have reached its natural end point: its fifty-five day and month will last until both worlds are destroyed in the Sun's death throes.

By that time, our descendants should have left for another planetary system, possibly by one of the spaceships envisaged in a later chapter. But the Moon has already proved to be a vital first step in Man's expansion out into the Universe. Some scientists have predicted that we could mine the Moon for its abundance of metals which are rare on Earth, such as titanium; perhaps artificial bases could be set up to act as homes on the bleak lunar surface. Such ideas may always remain pipe-dreams, for economically it may make more sense to mine the asteroids (see Chapter 6) and to live in wholly artificial colonies in space. But the nearness of the Moon to the

Earth has meant that Man has been able to reach out and set foot on another world in the very earliest years of the space-age – before far-simpler unmanned craft landed successfully on Mars or flew past Jupiter, Saturn or Mercury. The Apollo programme showed that Man *can* fly to other worlds, if the finance and the will are there. In centuries to come, Neil Armstrong's 'giant leap' from Earth to Moon will itself seem like a 'small step' – the first small step in Man's expedition to discover for himself this wide Universe around us.

— 4 —

Neighbouring Worlds

It's the most natural thing in the world to think that the Earth is the centre of the Universe. Our home seems firmly fixed, while everything moves around about us. Ancient astronomers thought the stars were points of light fixed to the inside of a huge black velvet dome, which gradually rotates about us so that stars rise in the east and set in the west during the course of a night. Closer to us than the dome of stars were the Sun and Moon. They take part in the majestic wheeling of the skies, once in twenty-four hours, but they were not fixed to the dome itself. The Moon moves slowly through the background of stars, returning to its original position after a month has passed. If we could see the background stars during the day, we would see the Sun moving too. Its more stately passage carries it once round the dome of the stars in a year.

The ancients naturally thought both the Moon and the Sun revolve about the Earth – the Moon being nearer because it moves more quickly and blots out the Sun during eclipses. In fact, of course, the Earth moves around the Sun during the year, not the Sun around the Earth. But it takes very precise measurements to prove this fact – the technique of measuring star distances by *parallax* (described in Chapter 11) relies on the Earth's yearly motion, but astronomers only looked for its subtle effects because they already believed that the Earth orbits the Sun. Men changed from believing in an Earth-centred Universe to a Sun-centred scheme of things not because of anything to do with the Sun itself, or with the stars or with the Moon; but because of the weird motions of five bright star-like points of light – the bright *planets*. If it were not for the planets – or the other planets, as we would say now that we know the Earth itself is a planet – astronomers would have taken much longer to dismiss the Earth from the centre of the Universe, and our understanding of the Universe would have been severely held up.

To the naked eye, the planets appear as bright points of light, like brilliant stars. They move slowly, so from week to week you can see that they shift amongst the constellation patterns of the stars. Mercury and Venus never move very far from the Sun in the sky: we see them either in the western sky after sunset as an Evening Star, or as a Morning Star in the east before sunrise. There, however, the resemblance ends. Venus is the most brilliant planet, shining ten times brighter than any star, and some-times looking like a small lantern hanging in the sky. Many reports of UFOs – flying saucers – are simply sightings of Venus by people who are not used to looking at the sky. If Venus is above the horizon, you just can't miss it. On the other hand, you are unlikely to spot Mercury unless you know exactly where and when to look. It never strays as far from the Sun as Venus, so Mercury is always low down in the bright twilight glow of sunrise and sunset, and in addition it is much fainter than Venus. Tradition says that the great Renaissance astronomer Copernicus, who changed many of the existing ideas about the Solar System, died without ever seeing Mercury.

The other three planets visible to the naked eye – Mars, Jupiter and Saturn – are not constrained to lie so close to the Sun. Their slow motions carry them right round the sky in a period of several years or decades. But unlike the faster movements of the Moon and Sun, the planets' gradual progressions are not smooth and steady. As Jupiter, for example, becomes prominent in the skies each year, it will slow down, stop, and then retrace its steps for four months. At the end of this time, Jupiter will apparently think again, slow down to a standstill, and turn to carry on its normal motion amongst the stars. The backward portion of the path is normally a bit 'above' or 'below' the forward track, so if we could see the motion speeded up Jupiter would seem to swing back on itself in a narrow loop.

These *retrograde loops* in the planet's paths were a major headache to the early astronomers who believed that everything revolves about the Earth. The Greek astronomer Ptolemy postulated that each planet was carried on the edge of a wheel, whose centre went around the Earth. By the Renaissance, astronomers found that this idea would not accurately de-scribe the planet's loopy paths: they were driven to put wheel upon wheel upon wheel until the whole arrangement was totally unwieldy. In 1543, the Polish monk Nicolaus Copernicus realised he could make sense of the planet's motions much better by assuming they go around the Sun, not the Earth. Then the Earth too must be a planet, following a path about the Sun. These ideas were heretical at the time, but they were confirmed in the early seventeenth century by the German mathematician Johannes Kepler. Kepler solved the remaining problems by showing that the planets' paths around the Sun are not circles, but egg-shaped curves called *ellipses*.

We now know that Kepler was entirely correct. Telescopes have revealed three more distant planets – Uranus, Neptune and Pluto – which also follow elliptical paths about the Sun. The nine planets are held in these orbits by the Sun's powerful gravitational pull. The orbits lie almost in the same plane, as if they are drawn on a flat sheet of paper with none of the orbits tilted much up or down. The Moon's orbit about the Earth lies almost in this plane too. From our viewpoint on Earth, everything in the plane of the Solar System appears to lie in a narrow band encircling the sky – a band we call the zodiac. The Moon moves around the zodiac once a month, the Sun makes its way steadily through a year, and the planets perform all their complicated motions within the band of the zodiac.

Once we know the Earth is a planet, the retrograde loops of the other planets are easy to understand – as Copernicus first realised. The Earth is moving faster than the outer planets like Jupiter. As our orbit brings us around to the side of the Sun where Jupiter happens to be, we overtake Jupiter on the inside, and so to us the other planet seems to go backwards. It's like travelling round on a fairground carousel, while a friend walks slowly around the outside in the same direction. When we are on the far side from the friend, we see him or her moving in the correct direction against the background of other stalls and sideshows; but as we come round to the same side, we sweep past the friend, and to us he seems to move backwards.

A telescope shows that the planets are not mere points of light, but worlds in their own right. We can see dark markings on the red deserts of Mars, and icy caps at its poles; swathes of clouds on Venus and Jupiter; and rings encircling distant Saturn. But there is only a limited amount we can discover about other worlds by merely looking at them. Just as a photograph of a celebrity tells us only a fraction of what that person is really like, so it needed spaceprobes to visit the planets before we could discover their individual personalities – and their family resemblances.

Although men haven't yet visited any other planet, automatic spacecraft have acted as remote eyes and arms to explore these worlds for us. The

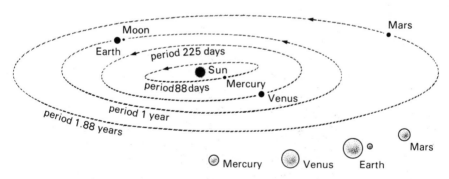

8 THE INNER PLANETS *The inner and outer planets of our solar system are strongly contrasting environments. The small, warm inner zone is occupied by the four 'terrestrial' planets, Mercury, Venus, Earth and Mars. Although superficially very different worlds, they have many broad similarities: they are all comparatively small and warm, made of rocky materials, with thin atmospheres, and they have only a few satellites if any. Mercury is about one-third the size of the Earth, only forty per cent larger than our Moon. Venus is almost the Earth's twin in size but Mars is also a small world, about half the size of the Earth. They lie so close to the Sun that their orbital periods are all less than two Earth-years in length.*

names of the American missions echo the romance of travel to other worlds: the comparatively simple craft that blazed the way to Jupiter and Saturn were part of the Pioneer series; their successors were the more sophisticated Voyager spacecraft, which have returned stunning photographs of Jupiter and Saturn and their moons. These latter were more advanced versions of the American workhorse, the Mariner probes. Earlier Mariners had returned the first close-up pictures of Earth's neighbours, Mercury, Venus and Mars. The last Mariner probe to Mars was followed by Viking spacecraft, which pillaged the soil of Mars in 1976 to search for signs of life. These spaceprobes – and Russian probes to Venus – have made the alien surfaces of other worlds as familiar to us as parts of the Earth that most of us will never visit in person: in the comfort of our living-rooms we can imagine the scorched deserts of Mercury or the iceball moons of Saturn in much the same way we visit in imagination the wastes of the Sahara or the Antarctic.

Mercury, the planet nearest the Sun, is a small world. At about one-third the size of the Earth, it is only forty per cent larger than our Moon. Earth-based telescopes show no more detail than vague dark markings on its surface, and we know the true appearance of Mercury only because the spaceprobe Mariner 10 flew past the planet in 1974. This spacecraft sent back the only close-up photographs of Mercury that we have. They show that Mercury is indeed a Moon-like body, completely devoid of an atmosphere, and with a surface completely pitted by craters of all sizes. Mercury has flat plains too, rather like the maria of the Moon, but bearing such a heavy load of cratering that they are not as easy to distinguish. It has one kind of feature that the Moon does not possess, though. These are long, winding ridges, some of them hundreds of kilometres long and reaching a height of 3000 metres (9800 feet) – similar to the Pyrenees on Earth. Unlike any kind of terrestrial or lunar mountains, the ridges of Mercury are unrelated to the surrounding terrain: they run through plains and craters alike, simply pushing up the existing features in long ribbonlike patterns.

Astronomers explain these ridges by hypothesising that Mercury's core has shrunk. Just as the skin of an old apple puckers up because it is too big for the shrunken flesh inside, so Mercury's solid crust is now too large for a core that has contracted since the crust first solidified. The idea that a planet can shrink is not as bizarre as it sounds. For a start, calculations show that Mercury needs only to have contracted by a couple of kilometres – less than a tenth of one per cent – to produce the observed ridges. In addition, Mercury has an unusually high density – for such a small world it is quite massive – and this means it cannot consist of rocky materials throughout. It must have a very large core consisting of denser matter, and scientists think that Mercury has a metal core consisting of an iron-nickel mixture, very like the Earth's core in composition, but filling a much larger proportion of the planet's interior. As Mercury has cooled down, some of the liquid metal in the core has solidified. Solid metal takes up less space than the same metal in liquid form, so Mercury's core has shrunk in size. The already-solid crust around has had to contract too, and the extra surface area has crumpled up into the long wrinkle ridges.

Although no spacecraft has landed on Mercury to test its rocks directly, astronomers think its history must have been broadly similar to the Moon's, up until the time when Mercury began to contract. Mercury is another fossil world, where all activity ceased at an even earlier date than it did on our Moon. It looks as though lava plains on Mercury flowed out while the planet was still under heavy bombardment from space, so they became almost as pitted with craters as the highland regions. Unlike the Moon, no new lava flowed after the impacts ceased about 3900 million years ago. The difference may be due to Mercury's shrinkage, which probably began about then. Then lava had flowed upwards, from deep reservoirs of molten rock, through cracks in the solid crust. As the crust started to contract around Mercury's shrinking core, the sides of the cracks would have closed up almost immediately and shut off the lava flows. Once all the cracks had closed, the crust began to pucker up into its characteristic ridges. And that is essentially the end of Mercury's history.

An explorer standing on Mercury's surface today would find it an even more hostile place than the Moon. The night-time temperature is similar – about $-170\,°C$ ($-274\,°F$) – but Mercury is so close to the Sun that the temperature at 'high noon' reaches $410\,°C$ ($770\,°F$) – hot enough to melt lead. Mercury rotates in such a way that the length of the 'Mercurian' day, from noon to noon, is twice as long as its 'year'! It completes one orbit around the Sun in 88 days (Earth-days), so this is Mercury's year. As seen from outside, Mercury turns round once on its axis in exactly two-thirds of this time, just under 59 days. But when Mercury has turned round once, it has gone two-thirds of the way round its orbit. If our Mercurian explorer was standing directly under the Sun 59 days ago, he is now facing the same direction in *space*, but because Mercury is in a different part of its orbit, the Sun is no longer in that direction. He must in fact wait 176 days (two Mercury years) before he finds himself once more facing the scorching heat of Mercury's mid-day Sun.

The next planet out from the Sun is almost Earth's twin in size. Unlike the Earth, however, Venus is perpetually veiled by a complete blanket of cloud, and as a result astronomers knew virtually nothing about Venus as a world until the 1960s. Some astronomers had thought Venus might be an Earth at an earlier phase of evolution: perhaps the clouds hid huge prehistoric rain forests where giant dragonflies fluttered round giant ferns, like those that have been fossilised into the coalbeds of our planet. The truth has turned out to be very different. Venus is a baking-hot world, hotter even than Mercury, where barren deserts lie under a heavy choking atmosphere capped by acid clouds.

Our first discoveries about the hostile planet beneath those all-enveloping clouds came not from spaceprobes, but from radio astronomers on Earth. Instead of just picking up the radio waves which are produced naturally in space, they sent out a powerful beam of radiation towards Venus, and then waited for some of the radio waves to return after reflection from the planet. Radio waves can penetrate clouds, so they expected a radio 'echo' from the hidden solid surface of Venus. Astronomers were for the first time actively probing outwards into space, using radio waves as elongated invisible fingers to touch another world. The technique is basically just a very sensitive form of the radar used by airports and ships to locate distant planes and vessels, and this active kind of radio astronomy is called radar astronomy.

By 1962, radar astronomers had equipment good enough to receive the weak reflected signal from Venus, some 41 million kilometres (25 million miles) away even at its closest. They could now measure its distance very precisely, simply by timing the delay between sending the signal and picking up the 'echo' (in a similar way to measuring the Moon's distance by reflecting laser light, described in the last chapter). A detailed analysis of the returning radio waves can however reveal much more about the target that reflected them. These early radar experiments turned up the first of a string of surprises about Venus: our neighbouring world rotates on its axis in the opposite direction to the others. Most of the planets rotate as the Earth does, west to east (anticlockwise if we look down on the Earth from above the north pole); but anyone who was stationed above Venus's north pole would see Venus turning clockwise. Even more surprising, it rotates very slowly, taking 243 Earth-days to make one complete turn.

More recently, radar astronomers have been able to make maps showing up mountains, valleys and craters on the hidden surface of Venus. The long fingers of radar beams from Earth cannot reveal the heights of Venus's landscape very accurately, however, and the real unveiling of Venus began in late 1978, when the Pioneer Venus spaceprobe went into orbit around the planet. It carried a small radar set that could map Venus at close quarters.

Stripped of its clouds, Venus appears as a very different world from the Earth. Our planet has two distinct levels, the ocean floors, and, standing some 4000 metres (13,000 feet) higher, the continents which cover about a third of the Earth's surface. Most of Venus, in contrast, is neither 'highland' nor 'lowland'. It consists of rolling plains, all very much of the same level, and covering two-thirds of the planet. The plains are broken by a few lower-lying basins, probably similar to ocean beds on Earth, and by three highland regions. One of these seems to be a huge double volcano, called Beta Regio. The other two are large plateaux, somewhat similar to the continental blocks of Earth, but rather smaller. Aphrodite Terra, lying along the equator of Venus, is roughly the same size as Africa; while Ishtar Terra, near the north pole, is only half as big. Ishtar is bounded by a steep mountain range, Maxwell Montes, which in places rises higher than our own Mount Everest.

Now that we have seen Venus's face unveiled, astronomers are in a much better position to understand the worlds of the inner Solar System. The Moon, Mercury and Mars are comparatively small worlds, and up until now the Earth has been the only large rocky planet available for study. Planetologists had expected that the interior workings of our twin-world Venus should be very similar to the Earth's. Although Venus would be dry, because of its searingly-high temperature, its surface should consist of distinct highland 'continent' regions and low-lying 'ocean-floor' basins, and these should be broken up into great plates of rock moving slowly around. At the plate boundaries there should be the long winding 'mid-ocean ridges' and deep troughs that we find on Earth.

Embarrassingly enough, Venus shows no indications that its rocky crust is composed of moving plates. And, as we have seen, most of the surface is neither 'continent' highlands nor 'ocean-bed' basins, but rolling plains at intermediate levels. On the other hand, Venus is not a dead fossil of the young solar system like the Moon or Mercury. It has volcanoes, which have probably been erupting until fairly recently – American planetologist Gordon Pettengill suggests the Beta Regio volcanoes may still be active. Some internal force must have pushed up the two big plateaux and their attendant mountain ranges, and broken open a few large cracks elsewhere on the surface. Geological activity – tectonics – is apparently going on, but on a much lesser scale than here on Earth. No one really knows why. A leading American expert on Venus, Harold Masursky, comments: 'There are two things which may have stopped the development of full plate tectonics; the lack of water or the slow rotation rate. But we have no proof that either one of these really has anything to do with plate tectonics.' Resolving this puzzle will tell us not only why Venus lacks 'full plate tectonics', but why the Earth is the only planet which *does* have a crust broken up into shifting plates, carrying around the continents on their backs – and so, amongst other things, affecting the evolution of life here.

The other differences between Venus and the Earth seem, at first sight, to be far more dramatic than the shape of its landscape. They include a surface temperature of 470 °C (878 °F) – hot enough to melt lead; an atmosphere a hundred times denser than the Earth's, and composed mainly of unbreathable carbon dioxide; and a cloud-deck of sulphuric acid droplets. If you stood on the night side of Venus, you would see the rocks literally glowing red-hot around you. The unbreathable atmosphere would press in on you with the same force that a submarine experiences 1000 metres (500 fathoms) deep in one of Earth's oceans. And a slow drizzle of concentrated sulphuric acid would fall from the sky. It's become something of an astronomer's adage that if you landed on Venus, you'd be simultaneously roasted, crushed, suffocated and corroded!

Both the Russians and the Americans have landed spaceprobes on the surface of Venus. Not surprisingly, none of them continued working for much more than an hour before their electronics succumbed to the heat and pressure. Imagine putting a transistor radio in a kitchen oven on its hottest setting and you will appreciate some of the problems: Venus is even hotter, and probes must bear the crushing atmospheric pressure too. The Russian probes Venera 9, 10, 13 and 14 actually managed to send photographs of their surroundings back to Earth, revealing a barren rocky plain stretching to the horizon. It's symptomatic of our drive to explore and understand the Universe that man should have sent remote eyes, ears and noses to a world that astronauts will never be able to visit in person.

The forbidding conditions on Venus all seem to result from the fact that it is slightly closer to the Sun than the Earth's orbit. In the early days, the two planets probably had very similar atmospheres, dense shrouds of water vapour and carbon dioxide spewed out by volcanoes. As we saw in Chapter 2, the Earth was cool enough for the water vapour to condense out as great oceans, and dissolve the bulk of the carbon dioxide. Plant life ultimately converted practically all the remaining carbon dioxide to oxygen. But Venus was nearer to the Sun, and the Sun's heat kept it always a little too hot for the water vapour to condense. Carbon dioxide and water vapour trap heat very effectively by the 'greenhouse effect' (described in Chapter 2), and so the atmosphere of Venus became hotter and hotter.

At some stage, most of the water vapour disappeared from the atmosphere of Venus. It probably rose above the heavier carbon dioxide, until it came out above the cloud layers, where direct sunlight broke the water molecules into individual atoms which escaped to space. The other, less abundant, gases from the volcanoes simply rose and reacted together to make the thick layers of acid clouds – chiefly sulphuric acid, formed from the choking sulphur dioxide familiar to all those who've visited volcanoes on Earth. The sulphurous droplets and the little remaining water vapour trap heat too, and help to give Venus's surface the record for 'the hottest place in the solar system' – with a temperature even higher than Mercury at noon.

Despite Venus's harsh climate, some scientists have speculated that men could one day live there. The trick would be to alter the planet's atmosphere to something like the Earth's, so that the vicious greenhouse effect would not work. The problem with Venus was that water never had a chance to condense, and the greenhouse effect ran away with itself. If we could move the Earth, as it is now, to the orbit of Venus, conditions would not be too hot for us to live quite comfortably, in the polar regions at least, were the Earth to become more cloud-covered and so reflect away more of the Sun's heat.

American astronomer Carl Sagan once suggested that our descendants may drop hardy bacteria into Venus's upper atmosphere, where they will photosynthesise the carbon dioxide to oxygen. As well as being suitable for us to breathe, oxygen is much less effective at trapping heat than carbon dioxide, so as the atmosphere gradually cleared, the temperature of Venus would fall. The Russian spaceprobes have however discovered that there's not enough water vapour left in Venus's atmosphere for this scenario to be feasible. Even if we could condense the vapour, it would not make rolling oceans – nor even a decent-sized sea. All the water in Venus's atmosphere would hardly fill one of America's Great Lakes, and in practice this water would immediately evaporate back into the dried-up atmosphere. Unless we can take our oceans with us, Venus will never become a second home for Man; future space-explorers may always give our red-hot, acid-swathed sister world a wide berth as they voyage outwards into space.

By contrast, the world beyond the Earth's orbit is a veritable paradise. Although Mars is a small world, about half the size of the Earth, its gravity is strong enough to retain a thin atmosphere. Mars is a colder world than Earth, because it lies further from the warming fires of the Sun; so cold that water is frozen to ice even in its equatorial regions. But temperatures here are no worse than those in our Antarctic. If Man ever sets up a second home within our solar system, it is likely to be on the orange-red deserts of Mars.

The surface of Mars is so strongly-tinted that the planet looks blood-red even to the naked-eye – which is the reason why ancient astronomers named it for the god of war. A telescope shows a more gentle, ochre-red little globe. At first glance, Mars looks disappointingly plain, but as your eye becomes used to uniform redness, you begin to see markings: white icy caps at its poles, and fuzzy greyish patterns mottling the surface. In the last century, astronomers believed that the darker markings could be huge tracts of vegetation, and some thought they saw long straight 'canals' bringing irrigation water from the melting polar caps across the deserts. A wealthy American, Percival Lowell, built an observatory especially to study the 'canals of Mars', and he mapped literally dozens of them, with 'oases' where the canals crossed. Lowell argued that such canals must be artificial, and so there must be intelligent life on Mars. The canals were so orderly that central planning must be at work, under a 'mind certainly of more comprehensiveness than that which presides over the various departments of our own public works'!

We now know that Lowell's canals do not exist: they were simply the result of the human eye joining up unrelated dark spots when it is straining at the limits of visibility. But if modern astronomy and spaceprobes have relegated Martians to the pages of science fiction, they have shown that Mars is a fascinating world in its own right. We know more about Mars than about any other world apart from the Earth and Moon. Although men have not yet travelled to Mars, a dozen spaceprobes have acted as remote eyes, ears, noses and arms to probe the planet. Their cameras have scrutinised the entire surface from orbit about the Red Planet, while lander probes have breathed the martian 'air' and listened to the light breezes. The American Viking landers even extended long arms with scoops to dig up the red desert soil for analysis. They acted under instructions from a 'brain' on Earth, via invisible 'nerves' of radio waves stretching 80 million kilometres (50 million miles) across space. Perhaps more than any other achievement in space, the Viking missions to Mars have shown how Man can reach out and touch other worlds far away in space. A laboratory technician must often handle radio-active substances behind safety glass with artificial arms and hands which act as an extension of his own, and the Viking scientists are in much the same situation: it makes little difference that their extended arms wave in the thin frozen atmosphere of another planet.

The eyes of the orbiting spacecraft have found that Mars is a curious cross between the Earth and the Moon, which perhaps isn't too surprising because Mars is intermediate in size between these two well-explored worlds. Like the Moon, Mars is divided into upland regions, heavily pitted with craters, and low-lying lava plains. Curiously enough, the two kinds of region aren't interspersed. Almost all the southern hemisphere is rugged, cratered uplands, while most of the northern hemisphere consists of the flatter plains pocked by only the occasional crater. Mars's early history must have been much like that of the Moon and Mercury. It had a solid crust which was bombarded by in-falling chunks of rock which blasted out the craters. After the worst of the bombardment was over, some 3900 million years ago, molten rock from small reservoirs below the surface flowed up and flooded the lower parts of the crust, in Mars's northern hemisphere, to produce the smooth plains. Mars's plains are the same ochre-red colour as the highlands, unlike the dark maria of our Moon which contrast so strongly with the bright highland regions. The reason is simple. Because Mars's gravity was strong enough to hold on to some of its atmosphere, the surface rocks originally lay under damp noxious vapours. As a result, Mars has rusted. The red deserts of Mars are coloured by compounds of iron, similar to the rust we find on an old nail that's been exposed to Earth's atmosphere. Both lava plains and highlands have corroded alike, to a uniform orange-red rust.

The dark markings that so intrigued early astronomers peering through their telescopes have nothing to do with the plains and mountains of Mars. They are just regions where the martian winds have blown away the brighter red dust, and exposed the darker rocks beneath. The most obvious features of Mars have turned out to be no more than wind-scoured areas, but the planet in fact harbours some of the most exciting geology in the solar system. Despite its small size, Mars has volcanoes far larger than any on Earth, and enormous canyons which put Arizona's Grand

Canyon to shame: the largest canyon could swallow the Alps, with room to spare.

Compared with the Moon or Mercury, Mars has been an active world until quite recently. The outflow of the great lava plains did not exhaust Mars's supply of molten rock. As Mars has cooled, lava from deeper underground pockets has welled up through single cracks to build up huge volcanoes. The largest of all is Olympus Mons (Mount Olympus), a giant of a mountain 25,000 metres (82,000 feet) tall and 400 kilometres (250 miles) across its base: this single mountain would cover Spain, and tower to almost three times the height of Mount Everest!

Olympus Mons is also the most recent of the martian volcanoes. Although it is not active now, its lava outpourings finished not too long ago on the geological timescale – perhaps only 200 million years ago. It is one of a group of youngish volcanoes which stand on a large plateau – pimples on a larger swelling which disfigures part of Mars's equatorial deserts. The swelling, called Tharsis, is 10,000 metres (32,800 feet) high itself and stretches over 4000 kilometres (2500 miles) of Mars: imagine North America raised to the height of Everest, and put on a world of half the Earth's size. Great fractures run down the sides of Tharsis, where the ground has split open as the enormous bulge has been pushed up. The most spectacular of these crack systems runs off to the east, almost along the line of Mars's equator. These huge, linked canyons are named the Valles Marineris (Valleys of the Mariner) after the Mariner 9 spaceprobe which startled astronomers with the first pictures of Mars's giant volcanoes and canyons in 1971. Even the smallest branches of the Valles Marineris would dwarf the Earth's Grand Canyon, and many of the huge rifts are hundreds of kilometres long, dozens of kilometres wide and 5000 metres (16,400 feet) deep. The entire network covers a region as large as the Mediterranean Sea.

Despite these monumental showpieces, Mars is not nearly as active a world as the Earth. Because it is smaller, its rocky interior has cooled more quickly, and geological processes have died down with the solidification of the interior rocks. In fact, Mars has never exhibited the Earth's most important geological process, the 'plate tectonics' which shift our continents about over hundreds of millions of years. The cracks around Tharsis show that Mars's crust *can* split, but it has never broken up into separate plates which move about the planet.

Geologists are however finding Mars tremendously useful: the features it displays are a half-way house to plate tectonics, which means we can study some of the processes of 'continental drift' without getting bogged down by all the details. The huge swelling of Tharsis may, for example, have been pushed up as the underlying rocks expanded when they solidified, or it may simply be a region of thin crust, where dozens of successful volcanoes have built up thick layers of solidified lava. Either way, it is telling us something about the behaviour of rocks that we can't learn on

Earth, where the underlying rocks have not yet frozen solid, and where it is not possible for successive volcanoes to build up huge deposits because the crust is not stationary.

The photographs from Mariner 9 surprised scientists not only with spectacular volcanoes and canyons, but with something much less obvious – but possibly even more significant. They showed dried-up river beds. If Mars has – or had – water, perhaps life could have evolved there.

There's one major problem, though. Liquid water can't exist on Mars. The planet is so cold that water would be frozen to ice: and even if Mars were somehow warmed up, its atmosphere is so thin that any pools of water would instantly evaporate into the martian 'air' whose pressure is only $\frac{1}{150}$ that of the Earth's atmosphere. On the other hand, there should once have been a lot of water on Mars. The martian atmosphere is made up mainly of carbon dioxide, with a little nitrogen and argon and traces of oxygen and carbon monoxide. Like the atmospheres of Venus and the Earth, these gases were originally belched out of the volcanoes of Mars's earliest days, and water vapour would have steamed off with them. This water should now be frozen into ice.

Some of it now lies in Mars's white polar caps, which are largely made of ice, although in mid-winter the temperature plummets to $-150\,°C$ ($-238\,°F$) and carbon dioxide gas freezes out on top of the ice caps to thicken them up. Much more ice is locked up in the soil. The surface of Mars is probably similar to the northernmost wastes of Siberia and Canada, where the ground is a permanently-frozen mixture of soil and ice, known as *permafrost*. In some places the martian permafrost has melted – possibly because of heat from underground lava streams – and here the surface has collapsed into a basin of jumbled boulders. The liquid water has gushed out of these basins in a sudden flood, and has left wide swirling channels, now dried-up. They look very like the Channelled Scablands of Washington State in the United States, which have been carved out when dammed-up water has suddenly escaped. On Mars, the channels come to a gradual end where the flowing water has evaporated away.

But the wide 'scabland' channels are not the only sign of running water on Mars. Up in the cratered highlands there are long, narrow, winding river-beds, complete with branching tributaries. These definitely aren't the work of a sudden flood, evaporating as it swept across the countryside. The sinuous channels must be the beds of rivers which kept flowing for years, if not centuries. They mean that Mars was once much warmer than today, and that its atmosphere was dense enough to prevent liquid water evaporating. These two things in fact go together, for a thicker atmosphere would have kept in more of the Sun's heat by the greenhouse effect, and so warmed up the planet.

We find the sinuous channels only in the martian highlands, which are the oldest remaining parts of the planet's surface. These rivers must have flowed only during the very earliest phase of Mars's existence. There's other evidence for a denser atmosphere at this time, around 4000 million years ago. Most of Mars's craters are sharp and crisp, because there's no running water now to wear them away, and the atmosphere is too thin to rub them away with blown dust or rock particles. But some of the oldest craters in the highlands have been worn down. Either water or a thicker atmosphere – or both – must have eroded these ancient features.

The polar regions of Mars contain another clue. Underneath the ice caps lie layer upon layer of clayey soil, possibly thousands of metres thick in all. During summertime, the ice evaporates to leave this surface exposed, and winds are stripping away the crumbly clay, digging deep valleys whose sides are lined with the different-coloured clay layers. The great deposits must have been eroded from rocks and swept up to the poles when Mars had a much thicker atmosphere than it does now.

The most likely explanation is that the early volcanoes of Mars belched out their gases in copious quantities very soon after Mars itself was born, so that the original atmosphere was something like one-tenth as thick as the Earth's. This 'air' kept Mars warm, and let the rivers flow; but the gases gradually escaped from Mars's weak gravitational pull. The atmosphere thinned to its present tenuous state, and the once-flowing water froze into the soil.

But influences from outside should have affected Mars too, just as they affect the Earth. The amount of sunlight could have changed, if the Sun is an inconstant source of light and heat, as some scientists believe (Chapter 7). We can't spot the signs of any such alterations on Mars with results from Mariner 9 or even from the much-better equipped Viking spaceprobes. But if future expeditions do find that the climate of Mars has varied in the same way as the Earth, then it will be proof that some outside influence is affecting both planets. On the other hand, it's possible that they'll find different variations on Mars from those that we find on Earth, and we shall then have to reluctantly abandon this theory. Such an investigation of Mars will be crucial in deciding what cosmic effects do in fact leave their mark on our vulnerable planet.

The layers of clays at the martian poles are already suggesting that our current theory to account for the ebb and flow of glaciations on Earth is largely correct. This Milankovitch theory (Chapter 2) explains the successive cold glacial periods and warmer interglacials as a result of regular changes in the Earth's orbit and in our planet's tilt. In the distant past, Mars's poles have suffered roughly regular fluctuations in climate. At times the winds would blow rock dust to the poles and deposit clay there; but then the climate would change for a while and the clay would no longer accumulate. Later, the climatic cycle would be completed and dust would be blown to the poles again, where it would form another clay layer. There's nothing on Mars itself to suggest why these climatic cycles should have kept on recurring, but the Milankovitch theory provides a very convincing solution. Mars's orbit is less circular than the Earth's, and its tilt can change more, and both of these factors would produce bigger effects than would occur on Earth. We won't know the answer for certain until we can measure the lengths of Mars's climatic cycles and compare them with the theory; but it's on the cards that continuation of the explanation of glaciations on our planet will in fact come from the frozen polar rocks of another world. Planetologist William Hartmann says 'the planet of "canals" and "Martians" has been replaced by a new Mars, just as interesting, which occupies a pivotal place in understanding our relationship to the solar system'.

For most scientists, though, the discovery of sinuous water channels spoke of something much more exciting than just climatic change. If there was once liquid water on Mars, then perhaps living things evolved there – just as living cells appeared in the early oceans of the Earth. Martian organisms could be lying dormant in the soil, ready to revive if the climate improves again. American scientists had this thought at the front of their minds when they sent the two Viking spacecraft to Mars in 1976. Each craft dropped a lander containing a miniaturised laboratory to look for signs of life in the red desert soil. Scientists on Earth manipulated their remote arms and sensors, confidently expecting each experiment to give a clear indication: 'life' or 'no life'. In fact, the results were confusing and contradictory. As Chapter 13 reveals, most scientists are now convinced that the martian soil confused the results because it contains substances which are very chemically reactive, and that there is no life on Mars.

But there is still room for hope. The Viking landers sampled only a few cubic centimetres of soil at two sites on the planet. Perhaps there are a few places elsewhere on Mars which are more suitable for living cells to have survived. American planetologist Robert Huguenin believes that there are some warm sheltered spots near the martian equator where ice can melt to water during the day. There may well be salts in the soil, too, which could act as a natural antifreeze and produce a briny mixture – rather like seawater – which would never freeze. It's certainly worth looking in places like these before declaring Mars to be a dead world.

Funding for American space programmes is unfortunately not as lavish in the 1980s, and most scientists would like to see what little money is available spent on exploring new territory – the world beneath Venus's clouds, the depths of Jupiter, or the strange satellites of the outer planets. The United States is planning no more Mars spaceprobes at the moment, though there are plenty of ideas on the drawing boards, including an unmanned wheeled vehicle, to travel up to 500 kilometres (300 miles) over Mars, and a robot aeroplane to fly over Mars at low altitude. But we can't fully understand Mars as a world until we have martian rocks and martian soil in a well-equipped laboratory, where we can measure their composition and age, and dissect them minutely for traces of life. If money became available, the United States could send an unmanned probe to Mars, to retrieve and return samples, by the end of the century.

The Russians have more grandiose plans in store. They have had problems with unmanned spaceprobes in the past – for example, only one of their seven Mars probes worked as planned – but the Russian programme of manned spaceflight has been building up in a slow but steady progression for two decades. They lost the 'space-race' to the Moon, with consequent loss of face. The Americans are not even in a race to send men to Mars, and the way is clear for the Russians to ignore the Moon and attempt to land the first man on Mars. And this isn't just speculation. A leading Soviet expert on space medicine, Oleg Gazenko, has recently said: 'It is difficult to give an exact date for a [manned] flight to Mars. But I think the basic prerequisites for such a flight exist now ... whether the flight happens in 10, 15, or 20 years, I cannot say. But I believe it will be before the year 2000.'

— 5 —

The Frozen Worlds

'Those incredible twisted rings defy the laws of orbital motion ... here we see the oldest surface in the solar system ... the youngest surface ... the smoothest surface ... the only other world with live volcanoes ... an atmosphere just like the Earth's original 'air', but preserved in deep freeze ... a world that's been split in two....' A string of superlatives and astonished exclamations pervaded the air at the Jet Propulsion Laboratory, California, in the years 1979 to 1981. Here scientists unscrambled the radio signals from the distant Voyager spacecraft, and turned them into pictures – photographs revealing for the first time fine details of giant Jupiter, ringed Saturn, and their amazing moons.

We can't tell much about the distant outer worlds of the solar system – Jupiter, Saturn, Uranus, Neptune and Pluto – with a telescope here on Earth. And their satellites appear merely as points of light, even though some are bigger then the planet Mercury. Two trail-blazing spacecraft, Pioneers 10 and 11, sent back some close-up pictures of Jupiter and Saturn in the 1970s, but these comparatively simple craft told us little that was new. When Voyagers 1 and 2 toured these giant planets and their moon systems, astronomers found themselves looking at totally unexpected kinds of worlds. To draw a parallel, imagine that you have spent your life gazing across a wide valley at an inaccessible house so far away that it appears only as a tiny speck in the sunshine; and then a magic carpet whisks you across to it so you can stand only an arm's length away, and inspect it so closely that you can see the sand-grains in the cement between the bricks. That's the kind of miraculous new view that the Voyagers gave astronomers as they sped past remote Saturn and its moons. These spacecraft have opened up the outer reaches of the solar system, and shown astronomers that the outer worlds are – if anything – even more interesting than our near neighbours, Venus and Mars. One team of American scientists said 'we would not have experienced a greater sense of novelty if we had been exploring a different solar system'.

The two Voyager probes were launched in the late summer of 1977, on powerful Titan–Centaur rockets that propelled them into space faster than a rifle bullet. Voyager 1 covered the 630 million kilometres (390 million miles) to Jupiter in only eighteen months, while its sistership took slightly longer. It so happens that Saturn, Uranus and Neptune all lie on the same side of the Sun as Jupiter in the early 1980s, and as a result the Voyagers were able to carry on outwards to intercept Saturn as well. The mission controllers in fact planned the probes' paths past Jupiter in such a way that they would pick up speed by robbing the giant planet of a minuscule fraction of its orbital energy. Thus Voyager 1 took only another twenty months to reach the mysterious ringed planet Saturn in late 1980. This remarkably successful probe is now heading off into the space between the stars, for Voyager 1 is travelling so fast that it will eventually escape from the Sun's gravity altogether.

Voyager 2 has a more interesting fate in store. Controllers on Earth arranged its flyby of Saturn in 1981 in such a way that the second Voyager's path will carry it on to meet the next planet Uranus, in 1986. If all is well, Voyager 2 will then proceed on to Neptune. Unfortunately, it cannot reach Pluto as well, but by 1989 this craft should have completed the Grand Tour of all four giant worlds of the solar system – Jupiter, Saturn, Uranus, and Neptune – and will be heading for interstellar space.

Like other space probes, the Voyagers carry our sense organs out into space: as well as carrying the camera 'eyes' which took those spectacular photographs, the probes 'listened in' to radio emissions from space, 'sniffed out' the kinds of gas which surrounded the planets in tenuous clouds, and 'felt' their magnetic fields. In addition, Voyager has a brain – 'perhaps the most sophisticated brain that's ever been put into space', according to Voyager scientist Bruce Murray. These probes had to be able to 'think' for themselves because they were so far away from Earth. When they passed Saturn, their radio signals spent an hour and a half on the long journey to Earth, and instructions took as long to get back to the probes. If the Voyagers had taken a picture and then awaited new orders from Earth, there would have been a gap of three hours between photographs. Instead, the controllers sent instructions weeks or months in advance to be memorised by the Voyager brain – or to be more precise, brains, for the Voyagers each carried three computers which were programmed to keep a check on each other. As a result, the probe could snap in quick succession some 17,000 photographs of each of the two giant worlds.

The huge world Jupiter is the colossus of the planets. It is as wide as eleven Earths, which means a hollow Jupiter could contain over a thousand planets like ours. And Jupiter is so massive that it would weigh more than twice as much as all the other planets put together. Some astronomers say we should really think of the solar system as being just the Sun and Jupiter, with all the other planets as insignificant afterthoughts. Jupiter weighs about one-thousandth as much as the Sun, and although it is by far the less important of the two bodies we should, strictly speaking, talk not of Jupiter orbiting the Sun, but of both bodies orbiting around their mutual balance point. Jupiter is in fact making the Sun swing about a balance point just

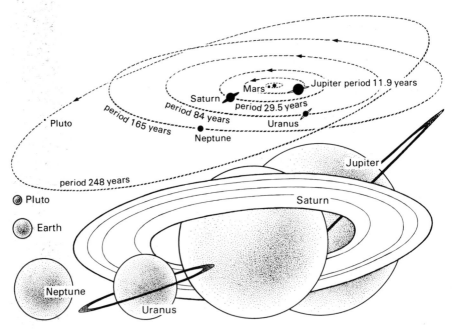

9 THE OUTER PLANETS *The outer solar system is a cold, empty region some thirteen times bigger than the zone of the terrestrial planets. It is occupied by four colossal gas giants – Jupiter, Saturn, Uranus and Neptune – whose enormous, freezing atmospheres, extensive ring systems (except in the case of Neptune) and vast satellite systems tell a great deal about the birth of our solar system. Their orbital periods are measured in tens and hundreds of Earth-years. Farthest out of all is tiny Pluto, a misfit world which takes 250 years to go once around the Sun and whose elongated orbit brings it on occasion closer to the Sun than Neptune.*

outside its surface, completing one circuit of this imaginary fulcrum every 'Jupiter-year' – 11.9 Earth years. In the year 1982, all the other planets lie on the same side of the Sun as Jupiter, and some scaremongers warned that the combined gravitational effect of this 'great alignment' or 'Jupiter effect' will backfire on the Earth, with disastrous consequences – including a major earthquake in California. Fortunately this is one 'cosmic influence' which doesn't stand up to close scrutiny.

One version of the theory states that the gravitational pull of the other planets, added to Jupiter's, will perturb the Sun into producing an excess of sunspots. As we shall see in Chapter 7, the sunspots come and go over a cycle of eleven years, and they may cause minor changes in the Earth's weather. But sunspots don't trigger earthquakes. And Jupiter doesn't cause the sunspot cycle. As if to prove the latter point, the current cycle of sunspots reached its peak in 1980, and the Sun's activity is now declining. One of the authors of *The Jupiter Effect* has now denied his own theory, in fact, and declared it 'a folly of youth'.

Another version is even more far-fetched. It declares that the planets raise extra tides on the Earth itself, and these will trigger the calamity. But even if all the planets lined up perfectly, they would cause an extra tide which is less than a ten-thousandth of the average tide due to the Moon. When the Moon is at its closest to Earth – as it was on 24 September 1980 – it exerts a greater pull than the Moon and planets *together* will in 1982. Besides, the planets are not perfectly lined-up: they'll be fanned out over more than a right angle measured from the Sun. Such an arrangement recurs every 179 years in fact, and no world-wide catastrophes were recorded in 1803, 1624, 1445 or at any earlier dates in the 179-year cycle; 1982 will see no more than an average year's share of disasters.

Jupiter has little direct influence on us here, but its huge bulk certainly reminds us that our Earth is a fairly insignificant world. Jupiter is not merely a scaled-up version of our planet, though, made of rocks surrounding an iron core. It is made up mainly of hydrogen and helium. Although we know these substances as gases, Jupiter's strong gravitational inpull crushes them so that they behave like liquids; only in the thin outermost layer of the planet can they actually behave as gases. In this shallow 'atmosphere' – as thin as the skin on an apple – float layers of clouds, stretched out around the planet in dark and light bands.

The photographs from the Voyagers show these clouds in ceaseless activity. New clouds erupt upwards like the towering piles of thunder-clouds; the bands at different latitudes swirl past each other, sending ripples and swirling eddies along their flanks; and larger, sedate eddies turn more majestically. The most spectacular of these is the Great Red Spot – a rotating vortex in Jupiter's atmosphere that is three times larger than the Earth. The Great Red Spot is a whirlpool in reverse, its rotation drawing up gases from deep down below. These gases turn red as sunlight breaks up their molecules and releases phosphorus – the same substance that makes a match head red.

All this motion reflects a struggle between forces. Sunlight warms up the atmosphere from above, and heat also seeps upwards from Jupiter's core: both heat sources create rising and falling columns of gas. Jupiter is rotating tremendously fast, too, turning round once in less than ten hours. Its rotation draws the gas motions out into the belts we see, rather like the Earth's pattern of 'Trade Winds' and 'Roaring Forties' at different latitudes. Meteorologists are in fact fascinated by Jupiter's weather, for it gives valuable information on our own. British meteorologist Garry Hunt says 'studies of planetary atmospheres are essential for understanding the meteorology and climate of the Earth ... One of the most exciting things to come out of the Voyager missions is that the Earth and Jupiter are very similar in terms of their weather systems.' Space probe missions may eventually help to provide one of the most important practical items in our everyday lives – accurate weather forecasts!

If we could peel off Jupiter's cloudy atmosphere, we would come across a thick ocean of liquid hydrogen and helium. This occupies almost all of Jupiter's bulk, apart possibly from a small core of rock right at its centre. About half-way down to the planet's core, we'd find the liquid so compressed that the hydrogen behaves like a molten metal. Electric currents flow through this metallic hydrogen, and generate a magnetic field, much as the Earth's magnetism arises in its molten iron-nickel core. As befits its giant status, Jupiter's field is 20,000 times more powerful than the Earth's. The invisible magnetic force reaches out into the surrounding space to fill a volume larger than the Sun. This huge magnetic field deflects the electrically-charged particles of the solar wind, keeping them out of this region, and out of a long empty 'tail' which stretches through interplanetary space as far as the next planet Saturn.

The Voyagers' most spectacular discoveries in Jupiter's neighbourhood however were nothing to do with the giant planet itself. First there was Jupiter's ring. Earth-based telescopes show that Saturn has several rings surrounding its equator out to several times the planet's own diameter, and indirect observation had shown that Uranus too has a set of very faint rings (see p. 65). According to the most widely-accepted theories of planet formation, however, the largest planet should have been fairly hot when it condensed from the original solar system nebula, and it would have boiled away any matter making up rings around Jupiter. But as Voyager scientist Bradford Smith put it, the team decided to look 'really for the sake of completeness ... you might as well look for a ring even if you don't think it's possible'. And, to everyone's surprise a ring appeared on the photographs. It's only a pale shadow of Saturn's spectacular appendages, though, and so faint it can't be seen from Earth.

Then there were the four large satellites of Jupiter, called the Galilean satellites because Galileo discovered them when he first turned a telescope to Jupiter in 1609. Unlike Jupiter itself, these moons are solid bodies with virtually no atmosphere, and they bear a strong family resemblance to the inner rocky planets like Mercury and Mars. Indeed, although they are technically 'satellites' or 'moons' because they orbit Jupiter rather than the Sun, these worlds are every bit as big as the smallest planets. The main difference in their make-up occurs because they lie so far from the Sun that water is frozen, and much of the satellites' solid bulk is ice rather than rock.

The Voyagers discovered that each of the four Galilean moons has an individual character. The outermost, Callisto, is Mercury's twin in size, though its ice and rock composition makes it much lighter in weight than the iron-and-rock Mercury. Callisto is the most heavily-cratered world ever seen. Jumbles of circular scars are littered all over it, one overlapping another, so that every square metre of its surface is part of one crater or another. All the solid worlds must have looked like this once, as they swept up the remaining chunks of rock which built them up (as detailed in the next chapter). On the others, however, flows of lava have obliterated some of the original densely-cratered terrain. The scarred face of Callisto appears to be the most ancient unchanged surface in the solar system.

Ganymede, second from the outside, is the largest moon in the solar system, bigger than Pluto and Mercury, and three-quarters the size of Mars. Some parts of Ganymede are heavily-cratered, like the surface of its sister Callisto, but these are isolated patches. They are separated by a strange kind of terrain – parallel lines of ridges, separated by deep cracks or grooves. In some places, as many as twenty ridges and grooves lie between patches of old cratered surface. Although they look small on the Voyager photographs, like cat scratches on a rubber ball, they are in fact immense: parallel mountain chains, each hundreds of kilometres long, ten to fifteen kilometres (six to nine miles) wide and 1000 metres (3280 feet) high. One Voyager scientist likens them   the ridges found in the Appalachian Mountains of the eastern United States. Ganymede seems to have suffered a simple kind of 'continental drift' long ago, with the grooved terrain being left as a stretch-mark where blocks of old cratered surface have drifted apart. Earth's continents float on a 'slush layer' of semi-liquid rock; but Ganymede is too cold for this. Its crust once floated on a much more familiar kind of slush – half-melted ice mixed with rocky gravel. Geologists believe that Ganymede – like Mars and Venus – will add some clues to the remaining problems of understanding continental drift on Earth.

Closer in to Jupiter, we come across Europa. Although the smallest of the Galilean satellites, Europa is still a substantial world, just a little smaller than our Moon. And it looks completely different from any other world known. Europa is a dazzlingly white globe, and it is completely smooth. It has no mountains, no grooves, and virtually no craters. Voyager scientists like to compare Europa to a billiard ball – except that Europa is if anything a bit smoother. Crossing this pristine white surface are crazy patterns of narrow light-grey lines, as if a small child had taken a felt-tip pen to the white billiard ball. These lines don't seem to be either ridges or grooves, but just dark markings in the surface: to pursue the analogy, the thickness of the ink would give the billiard ball lines a greater relief than that of the lines on Europa.

Many scientists believe Europa acquired its unique look because it warmed up slightly more than Callisto or Ganymede, up to a temperature where the ice frozen into the rocks could melt into liquid water. This rose to the surface, to jacket Europa in an all-enveloping deep ocean. Exposed to the chill of space, the surface waters froze into huge, brilliant-white ice floes, separated by narrow, darker channels of muddy water. Later, as Europa cooled, these channels also froze, to form the network of lines we see today.

These three worlds led the Voyager scientists to coin the phrase 'there's no such thing as a boring Galilean satellite'. But the innermost of these worlds turned out to be literally amazing. Little Io, only a fraction larger than our Moon, is the most active world in the solar system. It is a world of volcanoes. At any one time, at least half a dozen will be erupting, spewing sulphurous gases hundreds of kilometres high. Molten sulphur streams out from the volcanic vents, to coat its surface hundreds of metres deep, in glorious hues of yellow, orange and red. The entire world is covered with volcanoes' sulphurous vomit, and the volcanoes' throats between them take up fully one-twentieth of Io's surface. Io competes with red-hot Venus as the Hell-world of the solar system.

Astronomers had expected Io to be a little unusual, even before Voyager 1 photographed it in March 1979, because they knew it to be the reddest world in the solar system – redder than Mars – and suspected the colour was due to sulphur. Three American scientists, Stanton Peale, Patrick Cassen and Ray Reynolds, even published an amazingly accurate prediction in the journal *Science*: 'one might speculate that widespread and recurrent volcanism would occur [on Io]'. This appeared in print just three days before Voyager turned its cameras towards the red world.

The first photographs showed a blotchy red disc, 'like a pizza' or 'a rotting orange'. But as Voyager skimmed over its surface, scientists were particularly surprised by what Io does *not* have: craters. Even smooth Europa has a few, small shallow craters: even our planet Earth, with its powerful sandpapering forces of erosion, has a few dozen recent craters that are easily spotted from space. Io literally has none. Part of the reason soon became apparent, as the Voyager scientists spotted the long streams of molten sulphur which have evidently buried Io's original cratered surface deep down. It looked as though liquid sulphur – or possibly molten rock coloured with sulphur – was seeping out from Io's interior. No one seriously expected to see real volcanoes in action.

The discovery was largely accidental, though like most 'accidental' discoveries in science it was not pure luck that threw up the unexpected find, but a painstaking analysis of results acquired for a different purpose. Linda Morabito was in fact not an astronomer, but an engineer, when she found the volcanoes of Io. She was keeping track of Voyager 1's exact path in space by measuring the positions of the satellites in each photograph relative to the background stars. Off the edge of Io in one particular picture, she expected to see two faint stars: she turned up the contrast on the computer-enhanced picture – and she saw a very faint crescent hanging slightly above Io's bright rim. Instead of going on to measure the stars, Morabito decided to check up on this 'anomalous presence'.

She first checked that it wasn't another moon of Jupiter, lying in the background of the picture. She then 'contacted a camera expert who informed me that no known quality of the camera could [spuriously] induce the appearance of this anomaly'. Having decided it was really something to do with Io, Morabito measured its position, and found it was a huge plume lying directly over one of the great sulphur wells already identified on Io.

She concluded that she must be seeing an active volcanic eruption, in which sulphur and sulphur dioxide were being shot 300 kilometres (190 miles) out into space, and falling back in a wide umbrella-shaped 'plume' – 'the first ever witnessed on a body other than the Earth'.

After this discovery, the scientists checked the other Voyager 1 photographs and found eight volcanoes all erupting simultaneously. Seven of these were still in action when Voyager 2 passed by months later, though one had quietened down. The volcanoes bring up sulphur and molten rock from Io's interior, and they spew out enough sulphur in the liquid flows and the giant plumes to cover Io's surface with a fresh layer one millimetre (.04 inch) thick every year. This may not sound much, but it mounts up over the billions of years since the formation of the solar system. In this time, volcanic eruptions must have virtually turned Io inside out!

Io clearly flaunts the rule that only the larger rocky worlds – like Venus and the Earth – can have hot, molten interiors. Its internal heat therefore cannot come from the energy of radioactive atoms, as the Earth's does. What melts Io's rocks seems to be a gravitational tug-of-war. Io always turns the same face to Jupiter, just as our Moon always presents its 'Man-in-the-Moon' face to the Earth. It's in fact pulled out in a slight rugby-ball shape, with the length of the ball pointing at Jupiter. Unlike our Moon, though, Io also feels the gravitational tug of another nearby world – the next-out satellite, Europa. Europa's gravity tries to line up Io's long axis in its direction. Although Europa's pull is far weaker than Jupiter's, it repeats regularly as their orbits periodically bring Io and Europa close together. The recurrent strains on Io heat up and melt its interior rocks.

After the surprises of Jupiter, the Voyager scientists expected to be startled when the probes laid bare the secrets of Saturn and its satellites – and its famous rings. They were not disappointed.

Saturn is the second largest planet: although less than a third of Jupiter's mass, Saturn is still much heavier than all the others combined. Its interior is much the same as its big brother: a small rocky core, surrounded by compressed metallic hydrogen which generates a strong magnetic field, and, further out, ordinary liquid hydrogen and helium, surmounted by a thin, cloud-laden atmosphere. Saturn is a much blander-looking world than Jupiter, for its cloud patterns are obscured by a thick layer of haze. When scientists computer enhanced the plain yellow photographs from the Voyagers, though, they were able to pick out the patterns below – including a red spot, much smaller than Jupiter's Great Red Spot, but still as wide as the Earth. The photographs show that Saturn's clouds are swept around the planet by fierce winds. At the equator, the winds are considerably stronger than Jupiter's, and with a speed of 1800 kilometres (1100 miles) per hour they are ten times faster than 'hurricane-strength' winds on the Earth.

Most people, though – astronomers and non-scientists alike – were not as interested in Saturn as in its rings. A moderate telescope on Earth shows

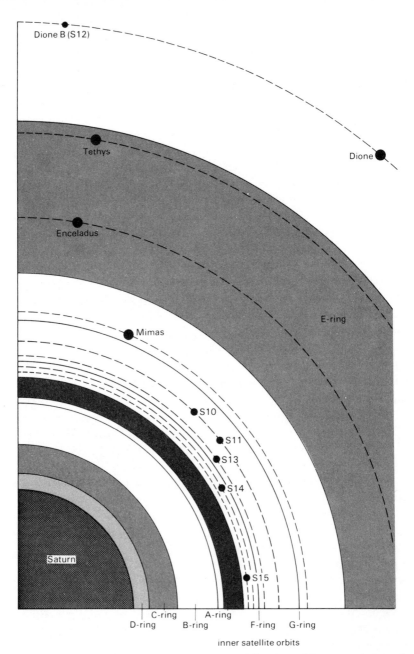

Dione B (S12)

Tethys

Dione

Enceladus

E-ring

Mimas

S10

S11

S13

S14

Saturn

S15

| C-ring | A-ring |
| D-ring | B-ring |
F-ring   G-ring

inner satellite orbits

10 THE RINGS AND INNER SATELLITES OF SATURN *The rings of Saturn are intimately linked with the innermost of its moons. The bright, wide rings A, B and C are easily seen from Earth; the D and E rings are much fainter. Spaceprobes discovered the very narrow F and G rings as they passed Saturn at close quarters. Satellite S15 controls the outer edge of the A ring, while S14 and S13 are 'shepherds' confining the narrow F ring. S10 and S11 are apparently the two halves of a disrupted satellite. Dione B is a small moon sharing Dione's orbit; Voyager 2 found similar tiny chunks of icy rock in the orbit of Tethys. Farther out lie Saturn's other large satellites, Rhea, Titan, Hyperion, Iapetus and Phoebe.*

us the wide band of Saturn's rings girdled around its equator, and reveals that there are in fact three broad concentric rings. The middle, B-ring, is the brightest. Separated by a dark gap to the outside is the A-ring, while the much fainter C-ring – the 'crepe ring' lies on the inside. Astronomers have suspected a couple of other rings, too, and in 1979 the simple space-probe Pioneer 11 detected a very thin outer ring, the F-ring, encircling Saturn outside the A-ring.

Even before the Voyager probes reached Saturn, astronomers knew that the rings are not continuous, like a sheet of paper, but consist of millions and millions of small chunks of ice and rock orbiting the planet like individual miniature moons. The smallest are the size of pebbles or bricks, while some of the larger boulders and 'icebergs' are probably as big as office blocks. The Voyager cameras could not photograph the individual chunks, but their pictures showed Saturn's rings have a structure that is stranger than anyone had imagined – even in the pages of science fiction.

The Voyagers discovered that the three main broad rings are in fact split up into thousands upon thousands of narrow 'ringlets', separated by equally narrow dark gaps. A close-up photograph of the rings looks just like the surface of a gramophone record, with its alternating narrow bright and dark circles (except here each is a complete circle, unlike a record whose groove spirals in towards the centre). Evidently, the individual rocky and icy chunks are channelled into certain particular circular paths. No one really knows why, though it may simply be that the thousand or so largest fragments in the rings are directing the paths of the others.

Another total surprise was the discovery of 'spokes' in the rings. Voyager 1 photographed broad bands that stretched outwards across the bright B-ring, and time-lapse photography showed that they rotate around the planet as if they are spokes in a 'wheel' formed by the rings. A very simple observation showed that the spokes consist of very tiny dust particles, rather than icy or rocky chunks. A large object, like a brick, always looks brightest when the Sun is shining full on it; if you look at the other side, it's dark and in shadow. Very small particles behave differently. If you look at a cloud of cigarette smoke outdoors in full sunshine, it appears dark. But if you look towards the Sun, through the cigarette smoke, it looks bright – because small solid particles scatter light forwards, rather than reflecting it back. As Voyager 1 approached Saturn 'out of the Sun', the spokes were dark against the bright rings; after close approach, the spaceprobe's cameras swivelled to see Saturn against the Sun – and saw brilliantly-shining spokes against the dark background of the rings. Thus the spokes must simply consist of the very smallest, microscopic particles in the rings. Astronomers aren't sure what bunches them together into spoke-shaped groups, however, although Saturn's magnetic field may be the culprit.

As these pictures flooded back to the Jet Propulsion Laboratory in California, the jubilant astronomers found themselves working around the clock to explain each new discovery – not just for their own sakes, but also for the hundreds of journalists and TV reporters who were eagerly sending the news around the world. But Saturn's rings had more in store. Voyager 1 was programmed to photograph in detail the narrow F-ring that Pioneer 11 had discovered the year before. The betting was that it would simply be another ringlet, but further out. Voyager's photograph was astounding. The F-ring is not just one ringlet, but three – and two of these narrow ringlets are braided together. One Voyager scientist put his feelings this way. 'When you've been working flat out for sixteen hours, had a few hours sleep, and are then confronted by something like that – it's rather tough on the mind!'

The chunks of rock and ice making a ringlet should be following smooth orbits around Saturn – either circles or egg-shaped ellipses. So two ringlets shouldn't twist in and out of one another over and over again. When Bradford Smith announced the discovery of the F-ring, he said it defies the laws of simple orbital mechanics – a phrase which was unfortunately misquoted on TV screens throughout the world, and startled viewers were informed that the F-ring defies the laws of science! The bizarre F-ring is telling us that some other force is affecting the chunks constituting it, in addition to the gravity of Saturn.

A likely explanation soon came up, when Voyager discovered two new moons of Saturn – both too small to be seen from Earth. One of these orbits Saturn just inside the F-ring, the other just outside. Theorists had already calculated that a pair of moons could confine tiny particles to make up a narrow ring between their two orbits: the F-ring is 'shepherded' between the newly-discovered 'sheepdog' satellites in exactly this way.

The discovery of the braiding stimulated theorists into action once more, and Stan Dermott in New York showed that the gravity of the 'sheepdogs' could in fact split the ring particles into two separate streams, which would intersect in 'braids'. Another force which could be at work is Saturn's magnetic field, though, and some astronomers believe that the braids could instead be a magnetic effect, causing ripples in a ring held captive by the 'sheepdog' satellites.

Voyager 1 discovered a third new moon for Saturn, close in, just outside the edge of the outermost of the three main rings. Its photographs also revealed that Saturn has two moons further out which move in almost the same orbit. Both are small, irregular chunks, and they are almost certainly the remains of a single moon that's been smashed in two by a tremendous collision. Two of Saturn's larger moons have narrowly escaped this fate. The moons Mimas and Tethys, visible as faint specks from Earth, both turned out to be disfigured by craters one-third the size of the moons themselves. With this great circular mark on one side, Mimas in particular looks like a huge disembodied eyeball gazing out into space.

Saturn's moons are great snowballs, ranging from ten kilometres (six miles) to 1500 kilometres (930 miles) in diameter. All of them bear craters, attesting to the final bombardment in which they were born, but they all have different characters. Hyperion is a large, flattened moon – 'like a hamburger', according to Voyager scientists: while another outer moon Iapetus looks like a scoop of ice cream sprinkled with plain chocolate powder. No one knows where this dark coating has come from, but it gives one half of Iapetus the darkest surface of any body in the solar system. Enceladus has a grooved surface, rather like Jupiter's moon Ganymede, though it's only one-tenth of the size. These moons are made up mainly of ice, with only a little rock mixed in, and that makes them a new kind of object to study. Planetologist Harold Masursky says 'we're now getting to the Antarctic equivalent in the exploration of the solar system, and the whole game is different'.

But the most interesting of all Saturn's moons is the largest, great Titan. Before Voyager 1's close encounter, astronomers thought Titan to be the largest moon in the solar system, but the spaceprobe pictures showed that it is slightly smaller than Jupiter's Ganymede – although still larger than the planet Mercury. Titan does however have a unique characteristic, much more interesting than mere size. It is the only satellite known to have a substantial atmosphere. Earth-based astronomers have had indirect evidence for Titan's atmosphere for nearly forty years, and the Voyager scientists put Titan at the top of their list of priorities for investigation, ranking with Saturn itself.

Voyager 1 investigated Titan thoroughly, with all the instruments on board. It discovered that Titan does indeed have an atmosphere – twice as dense as the Earth's, and composed mainly of nitrogen. There's a small amount of the simplest organic gas methane ('natural gas' on Earth), along with some ethane, acetylene and hydrogen cyanide. These substances particularly excited the scientists, because Titan's atmosphere has turned out to resemble our best guesses at what the Earth's original atmosphere contained – once we allow for the fact that Titan is so cold that all the water vapour and carbon dioxide must be frozen out, and make up parts of Titan's solid bulk rather than its atmosphere.

Unfortunately, we still know nothing for certain about Titan's surface. Voyager 1 skimmed past this world so closely that it could have shown up details as small as an ocean liner – but all it saw were clouds, clouds and more clouds. Titan is completely wrapped up in inpenetrable layers of orange cloud – an outer-solar-system version of Venus. Frustrating though it was, Titan's cloud cover is one of the most interesting phenomena in the solar system, and one of the most important to us. The clouds undoubtedly consist of complex organic molecules welded together from the simpler gases of Titan's atmosphere; preserved in Titan's deep freeze, they may tell us how the first molecules of life were assembled on the early Earth (see Chapter 13).

Voyager's radio measurements showed that Titan's hidden surface is frozen to a temperature of $-180\,°C$ ($-292\,°F$). In these frigid conditions, methane can liquefy, and just about solidify – in other words, it will play the role that water does on Earth. Under Titan's opaque orange sky, low banks of cloud drizzle a thin rain of liquid natural gas on to a landscape sculpted from ice, where it flows in rivers of liquid methane to wide oceans of methane, which may cover most of Titan. At its poles, great white caps of solid natural gas may calve off 'methane-bergs' into the warmer seas around. And some scientists speculate that the oceans may be topped with a thick slime of organic 'gunge', the greasy precursor molecules of life, that have dripped down from the orange cloud-deck far above.

This travellogue on Titan's surface is largely speculation of course, but by concealing itself so well Titan has made itself even more alluring to astronomers. It is certainly many scientists' top priority in a list of worlds to be further explored – when the money is available for a new (and certainly expensive) spaceprobe mission.

Meanwhile, Voyager 2 is on its way to the next planet. Although Uranus is another giant world, four times the size of the Earth, it lies so far from the Sun that it can only just be seen with the unaided eye – and in fact no one noticed this dim, slow-moving point of light until only two hundred years ago. It was another of those 'accidental' discoveries. A Prussian musician living in England, William Herschel, had taken up an amateur interest in astronomy, and because he could not afford to buy a good telescope, he started making his own. Herschel's home-made telescopes could not be pointed as precisely as the professionals' instruments, but they were made to be used with much higher magnifications. Herschel began to inspect all the stars in the sky under high magnification to find out which ones were in fact a close pair of stars; one evening in March 1781 he found a 'star' that wasn't a mere point of light, but appeared as a tiny greenish disc. Over the following nights, it moved slowly across the sky. Herschel's patient, tedious search had turned up something far more interesting than just another double star: he had discovered a new planet in our solar system, now known as Uranus.

Herschel's planet has one outstanding oddity. Its axis of rotation is tipped up so that it lies almost in the plane of its orbit around the Sun. Most of the other planets are slightly tilted over and, as we saw in Chapter 2, this produces the seasons. During the course of Uranus's 'year', 84 Earth-years in length, it must have a bizarre cycle of seasons. During the 1980s, one of Uranus's poles is pointing more-or-less directly towards the Sun. This polar region must in fact be warmer than Uranus's 'tropical' regions around its equator! The other pole is in the middle of a long polar night. It's been turned away from the Sun for 21 Earth-years, and another 21 years will pass before Uranus has gone far enough around its orbit for the Sun to begin to shine on this frozen pole.

Uranus has five satellites, and a strange system of dark narrow rings, found in 1977. For once, this discovery was not made by a far-flying multi-million dollar spaceprobe, but by a telescope on Earth – or to be accurate, sitting aboard an aeroplane above the Indian Ocean. In March of that year, Uranus would move in front of a distant star, and astronomers intended to study the star's disappearance and reappearance behind the planet's disc, for this would reveal some details of Uranus's atmosphere. Unfortunately, this event would only be seen from a small region of the Earth: West Australia, South Africa and the Indian Ocean. American astronomers however have a flying telescope, mounted in a converted C-141 transport plane. They took this Kuiper Airborne Observatory (named after the late Gerard P. Kuiper, a leading planetologist) down to the southern hemisphere to watch Uranus and the star.

They had to get the equipment set up well in advance, partly to check that it was working properly, and partly because there's always some uncertainty in the predicted time, it's better to switch on too early than too late. As they watched Uranus gradually creep up to the star, the astronomers were startled to see the star suddenly 'switch off'; then a moment later reappear. As they rechecked the equipment, the same thing happened again; and again . . . five times in all, before Uranus itself hid the star.

After the star reappeared behind the planet's disc, it repeated its sudden disappearing act another five times. There could only be one explanation. Uranus is surrounded by a set of five narrow rings, so densely packed with chunks of rock that they can block off a distant star's light as they move in front of it. Later observations have revealed that Uranus has at least nine of these rings, and all are very narrow – generally just a few kilometres across. In close-up, each would probably look like one of Saturn's myriad throng of narrow ringlets, perhaps like the famous braided F-ring of Saturn. Uranus's rings are far darker in colour, though. They can't be seen directly from Earth, even with large telescopes, so they must be as dark as coal – probably made of rocky rubble, rather than the bright ice chunks which populate Saturn's rings.

The last of the giant planets, Neptune, was discovered by mathematics. By the early eighteenth century, astronomers realised that the newly-discovered Uranus wasn't following a normal orbit about the Sun: it was evidently being pulled by the gravity of a more distant unknown world. In the 1840s, the English mathematician John Couch Adams and the Frenchman Urbain Leverrier calculated where this world should be; in September 1846 astronomers at the Berlin Observatory scrutinised this part of the sky – and there indeed was a new planet.

Neptune is very similar in size to Uranus. Both are probably built up in much the same way, with a central core of rocks surrounded by a thick layer of liquid water, mixed with ammonia and methane. On top, there's a cloudy atmosphere. Both worlds are so far off that it's difficult to see any details with Earth-based telescopes. Speaking at a meeting in early 1981 to commemorate the bicentenary of Herschel's discovery of Uranus, American astronomer John Caldwell described our view of Uranus from Earth as 'terrible'. He pointed out that if you were faced with ordinary photographs of such poor quality, 'you wouldn't be able to distinguish Margaret Thatcher from Lady Diana Spencer'.

Astronomers are keeping their fingers crossed that Voyager 2 will still be working as its path takes it past Uranus in January 1986, and Neptune in 1989. It will actually show us Uranus's rings for the first time, and photograph in close-up Neptune's huge moon Triton – probably only a shade smaller than Jupiter's Ganymede and Saturn's orange-coloured Titan. It will discover what the two distant giant planets really look like – and will, amongst other things, see how long they take to spin on their axes. Astronomers actually don't know how long the 'day' is on either Uranus or Neptune (though they are certain that the values quoted in older textbooks are wrong!)

Beyond the giant planets lies the orbit of tiny Pluto. Unlike the almost-circular paths of the other planets, Pluto's orbit is noticeably elongated – so much, in fact, that Pluto's path actually crosses Neptune's orbit. For twenty years out of its 250-year journey around the Sun, Pluto lies closer to the Sun than Neptune. This strange state of affairs is occurring at present. In 1979, Pluto crossed Neptune's orbit, as its path brought it closer to the Sun, and it will re-cross Neptune's track on its way out again in 1999. So, although Pluto's *average* distance is greater than Neptune's, at the moment Neptune is the farthest planet from the Sun.

The discovery of Pluto makes a strange story indeed. At first it was hailed as a triumph of the mathematicians, like the successful prediction of Neptune's position in the sky; but with hindsight the unearthing of Pluto has come to be seen merely as the product of a diligent hunt. By the end of the nineteenth century, astronomers were convinced that both Uranus and Neptune were being pulled by another unknown planet farther out still. Two Americans, W.H. Pickering and Percival Lowell, calculated where this planet should lie and their results agreed quite closely. Lowell had already built a private observatory at Flagstaff, Arizona, to study the 'canals' which he was convinced he could see crossing the deserts of Mars. Now Lowell added the search for 'Planet X' to the observatory's tasks.

But there was no planet as bright as expected near the predicted position. The search widened to include a larger area of sky, and a hunt for a fainter world. Success came only fourteen years after Lowell's death. Early in 1930, a young assistant, Clyde Tombaugh, found a faint speck of light in the constellation Gemini. It moved so slowly from night to night that it must lie beyond the orbit of Neptune; and, although much dimmer than expected, it lay reasonably near Lowell's predicted position.

Everyone was convinced that Tombaugh had indeed found Lowell's

Planet X. It's no coincidence that the first two letters of the name eventually chosen – Pluto – are the initials of Percival Lowell. But Lowell had predicted the new planet must be several times heavier than the Earth, to produce the gravitational pull on Uranus and Neptune which lay at the root of his calculations. We now know however that Pluto is a lightweight world, only $\frac{1}{500}$ as massive as the Earth. Pluto's feeble gravity cannot affect either of the two giant planets to any measurable extent, and modern calculations suggest that Uranus and Neptune are *not* suffering perturbations from any unknown body after all. So it has turned out to be only lucky chance that Pluto lay close to Lowell's predicted position. But even if it hadn't been there, Clyde Tombaugh would have eventually found Pluto. He was embarking on a diligent search of the whole band of the zodiac, and sooner or later he would have come across Pluto. Even after his discovery, he carried on the search so we now know that the solar system does not contain any more planets, unless they appear very much fainter even than dim Pluto.

Pluto is the smallest and least massive of all the planets. It's about three-quarters the size of Mercury, very much the twin of our Moon in size, but weighing only a quarter as much. Out here at the edge of the solar system, the Sun's rays bring hardly any warmth, and at a temperature of $-230\,°C$ ($-382\,°F$), Pluto is a ball of lightweight frozen methane. Many astronomers wonder whether we should really call Pluto a true planet at all, but perhaps regard it simply as an extremely large example of the asteroids or minor planets we shall meet in the next chapter.

In 1978, American astronomer James Christy found out another strange fact about this odd little world: it is actually a double planet. Christy discovered that photographs of Pluto sometimes seem to be rather elongated. He suggested, correctly as it's turned out, that this pear-shaped image is a blend of two images, both blurred by the Earth's atmosphere. The brighter image is Pluto itself; the fainter a previously-unknown satellite, which Christy has called Charon.

Other astronomers have subsequently found that Charon is a substantial world in its own right, one-third to one-half as large as Pluto itself. Planetologists now often regard the Earth and its one-quarter size Moon as so nearly equal that they should be called a double planet; now, in the very outermost fringes of the solar system, we have come across another double planet system in the frozen methane snowballs of Pluto and Charon.

Pluto's orbit marks the limit of our planetary system. There may be other planets beyond, but they must be dim – either insignificant worlds like Pluto, or very distant indeed. Tombaugh calculates that his search would have picked up any giant planet like Jupiter which should happen to orbit the Sun any closer than twelve times Pluto's immense distance. And there's little we could learn about such dim and distant worlds anyway.

Spaceprobes have taught us that we still have a wealth of exciting discoveries to make among the known worlds of the solar system. Before the first spaceprobes extended our senses to other worlds in the 1960s, astronomers were confident that they understood the planets – in general, at least. But, as British science writer Arthur C. Clarke puts it bluntly, 'pretty well all we thought we knew was nonsense'.

With probes like the long-distance Voyagers and the Mars-landing Vikings, we are exploring fresh lands. In the Renaissance years, men set out to explore our world, the Earth, and by the beginning of this century men had charted its secrets from the chilly poles to the hot humid rain forests of Africa and America. One day, men will know the solar system as well as they know the surface of the Earth. We, today, are luckier than we generally realise. We are living through the excitements of the second great Age of Exploration; the age when our craft are setting forth to discover what are literally New Worlds.

— 6 —
Birth of the Solar System

Five thousand million years ago, our solar system did not exist. The Milky Way Galaxy was already 10,000 million years old, and thronged with stars of all ages. Between the stars lay dark clouds of gas and dust, waiting their turn to leave obscurity and turn into blazing stars themselves. One of these clouds is particularly important to us – not because it was any different from its neighbours in space, but because this cloud happened to turn into the star we call the Sun. Its fringes became the planets of the solar system, including our home, the Earth.

A star is born when a cloud of gas and dust condenses down to a much smaller size, the condensation getting hotter until nuclear reactions start at its centre, and the newly-born star begins to shine in its own right (Chapter 8). But it takes quite a hefty kick to start a cloud on this road, to squeeze it sufficiently hard that it will continue to collapse in on itself. Our particular cloud showed no inclination to collapse until 4550 million years ago. At that time, a nearby star suddenly exploded in the holocaust of a supernova outburst (Chapter 8). The blast of gases from the explosion swept by our cloud. The shock of the impact squeezed the cloud so that its gas and dust occupied a much smaller region of space. In these denser conditions, the cloud broke up into a multitude of condensing fragments, each of which became a star.

We shall follow the story of one fragment of the cloud in detail; though no different from the others, it was the embryo of the Sun and our solar system.

This fragment shrank in size, drawn inwards by its own gravitation. But it was rotating, too. Centrifugal force prevented all the gas and dust from collapsing to the centre, and left a lot of it in a rapidly spinning disc, whirling about the central condensation which was to become the Sun.

In this disc – roughly the size of the present solar system – the tiny grains of dust found themselves much closer together than before. They now jostled one another as they whirled round in the disc, and as often as not they stuck together when they touched. The microscopic dust grains built up into solid, fist-sized chunks of matter. The tiny dust grains in space are thought to be mainly made out of rocky materials and ice (Chapter 11), so the first solid lumps of matter should have been a mixture of fine ice-crystals and specks of rock – just like dirty snowballs.

But even at this early stage, forces were at work which would forge the individual characters of the planets. Close in to the centre, hot condensing gases were building up the Sun. As a result, the inner regions of the great spinning disc were much warmer than its outskirts, which remained at the sub-zero temperature of interstellar space. In the outer reaches, then, the first solid matter would indeed have been these dirty snowballs. But closer in to the Sun the disc was so warm that the ice melted, then evaporated into water vapour. The first solid matter in the inner parts of the disc thus consisted only of rocky matter – pebbles in space. The water vapour joined the other gases – mainly hydrogen and helium – which filled the disc around and between the solid particles, as they followed their careering paths about the nascent Sun.

The pebbles in the inner reaches of the disc had now grown big enough to produce a weak gravitational pull on their neighbours. The great swirling carpet of equally-spread pebbles clumped together to make up rocky bodies a few kilometres to a couple of dozen kilometres in size. These are known as *planetesimals*, for they were the building blocks which eventually built up the planets.

As the planetesimals whirled about the Sun, they often collided: French astronomer Hubert Reeves describes the early solar system at this time as 'somewhat akin to a stock-car race'. The colliding planetesimals would often smash each other into small pieces; but sometimes their gravity would bind them together as a larger rocky body. Once one planetesimal had grown larger than its neighbours, it held all the trumps. This superplanetesimal had stronger gravity to pull in the other planetesimals; and it was so big that these impacts wouldn't break it up. Four or five such superplanetesimals gained the upper hand in the inner solar system. They swept up all the remaining planetesimals, and grew fat on their success: they became the rocky planets Mercury, Venus, Earth and Mars – and the Moon, if our satellite was originally a separate world (Chapter 3). After they grew to virtually their present sizes, these worlds took several hundred million years to sweep up the remaining planetesimals; and the impact of these has blasted out the craters we see today on the faces of the rocky worlds. In previous chapters, we have followed the subsequent adventures of these inner planets.

The dirty snowballs in the outer parts of the solar system also grouped together into planetesimals – huge icebergs in space – and these too grew into worlds. In the satellites of Jupiter and Saturn we see rock and ice worlds, formed from the iceberg planetesimals in much the same way as the inner planets grew from rocky planetesimals. These satellite worlds were saved from becoming part of the giant planets themselves, because they happen to lie in safe, almost circular orbits around the planet concerned. Most of the planetesimals in the outer part of the Sun's disc however were swept up by the four largest planetesimals; and each grew into a giant planet. Uranus and Neptune are just such rock and ice planets, although their interior heat has melted their original ingredients to create great liquid worlds.

Jupiter and Saturn fared differently. They were born in a region of the disc rich in gases around the planetesimals, and when they had grown to the size of Uranus and Neptune, their gravitational pull was strong enough to pull in these gases. Once they had started gorging themselves on the abundant gases, the whole process ran away with them. The extra gases made Jupiter and Saturn more massive; as a result, their gravitational pull became stronger and could pull in the gases even more. These planets managed to sweep up not only the planetesimals in their region of the disc, but the gas too. They grew and grew to become the hydrogen-helium giants that now dominate the solar system.

Astronomers have studied a whole range of clues in trying to understand the birth of the planets. They have fitted together each clue, like the pieces of a jigsaw puzzle, to produce the overall picture outlined above – although it's fair to say that the jigsaw still contains a lot of holes, and loose pieces still to be fitted in. On the one hand, radio and infrared astronomers are winkling out the secrets of starbirth in clouds where stars are being formed at the present day; on the other, theorists with powerful computers are calculating just what *should* happen in a whirling disc of dust and gases.

But the major witnesses are silent: the planets themselves tell us very little indeed about their birth. The surface rocks of the inner planets melted and changed from their original type; the prolonged sweeping-up of planetesimals kept pummelling the surface and altering it still further; while later geological changes may have mixed and changed the rocks again. The 'oldest rocks' on Earth took their present form only 3800 million years ago; rocks brought back from the ancient highlands of the Moon date back only 4200 million years – except for one particular rock lump, which has remained unchanged since the birth of the solar system 4550 million years ago.

Planetologists would dearly love to study rocks which have remained unchanged from the earliest days of the solar system. And, remarkably enough, Nature does provide us with such specimens – and free of charge. These are the *meteorites*, lumps of rock that fall to Earth from space. They are part of the material of a planet that never formed.

This planet would have circled the Sun between the orbits of Mars and Jupiter. Planetesimals started building up here, and grew into several dozen small worlds, each a few hundred kilometres across. But they never got together to make a planet – possibly because they were perpetually disturbed by the gravity of their huge neighbour Jupiter. Instead, collisions between these superplanetesimals started to break them up again, spilling rubble of all sizes into space.

This rubble still circles the Sun just beyond the orbit of Mars: these bodies are known as the *asteroids*, or minor planets. The largest asteroid, Ceres, is just over 1000 kilometres (620 miles) across – less than one-third the size of our Moon. Binoculars will show the brightest asteroids, like Ceres, if you know where to look. These tiny worlds show as star-like specks of light even in large telescopes, and only their motion across the sky gives them away. However, we may now have a closer view of what an asteroid looks like. The Viking spaceprobes photographed the small moons of Mars, Phobos and Deimos, which many scientists believe may be asteroids captured by Mars's gravity.

Astronomers have named the two thousand largest asteroids, and have stored their orbits in computers to keep track of them. As the size of asteroids being considered decreases, so their numbers increase. It's estimated that there are three-quarters of a million asteroids bigger than one kilometre (.6 mile) in diameter – no wonder some astronomers call them 'the vermin of space'! But despite their numbers, the total amount of matter in the asteroid belt is pretty small. If we could lump all the asteroids together, we'd end up with a minuscule planet, only one-thirtieth the mass of our Moon.

The original superplanetesimals in the asteroid zone heated up inside because of energy liberated by the radioactive elements within them, and the iron in the rocks melted out to make a small central core – much like the Earth's metallic core. Around this was molten rock, and covering the surface was a thin layer of the original rocky matter of the planetesimals. As collisions broke up some of the superplanetesimals, the debris formed three types of asteroids. Fragments of the iron core became metallic asteroids; pieces from farther out formed rocky asteroids with a composition like lava; while fragments of the original surface were once more liberated into space, still virtually unchanged since the birth of the solar system. These last asteroids are dark in colour, because they contain more carbon than most, and astronomers call them the *carbonaceous* type.

If these little worlds never strayed from their home 'asteroid belt' between Mars and Jupiter, they would be of less interest to us. But occasionally, asteroids are deflected on an inward path that crosses the orbit of Mars and the Earth, and some of them will collide with the Earth, to land as meteorites. In this way, astronomers have been able to build up collections of extraterrestrial rocks.

Meteorites are much more interesting than the much-publicised Moon-rocks, which many people queued for hours to see. The lunar rocks came from just the surface of one world; meteorites are fragments from many worlds, culled at all depths – from iron meteorites at the original core, through stony meteorites, to the carbonaceous meteorites from the original surface. And scientists take particular pains to study the latter, because carbonaceous meteorites are the oldest unchanged rocks we can lay our hands on: they are the solar system's birth certificate.

The carbonaceous meteorites tell astronomers what the original 'rocky' matter of the solar system was like. But more important, radioactive atoms in the rock provide an accurate 'clock'. Uranium atoms, for example, degrade slowly into inert lead atoms. By comparing the amount of lead that's been produced with the amount of uranium still remaining, scientists can tell how 'old' a rock is. For the rocks of the Earth or Moon, this is the time since it was last melted, or disintegrated by infalling meteorites. But in the case of the unaltered carbonaceous meteorites, the uranium/lead clock is telling us when the rock first formed, when it condensed from microscopic grains. And all carbonaceous meteorites agree that this event – the formation of the solar system – was 4550 million years ago.

The meteorites also indicate that their parent bodies, and presumably the other planets, all built up pretty soon after this – 'soon' on the astronomical time-scale meaning less than a hundred million years. Modern techniques for prying open the meteorites' secrets are revealing in finer and finer detail the precise events that occurred during this time.

Many of the new instruments for analysing meteorites were developed for studying the Moon-rocks brought back by the Apollo astronauts. For this reason, one of the leading meteorite laboratories in the world, at the California Institute of Technology, is officially called The Lunatic Asylum! The instruments in such laboratories are incredibly sensitive. An ion microprobe can concentrate energy on a tiny spot on the cut surface of a meteorite, and vaporise a microscopic amount of the mineral there. The experimenter can focus the microprobe on a spot only one-hundredth of a millimetre in diameter, so he knows he's studying the atoms from just one individual crystal in the rock. The tiny puff of vapour from the rock passes into the vacuum chamber of a mass spectrometer, where electric and magnetic fields sort out the individual atoms, and electronic detectors count up the numbers of each type. The mass spectrometer not only separates the different elements from one another – say carbon from iron – but can also distinguish between the very slightly different types of atom (called *isotopes*) that make up any particular element, for example, the relatively stable uranium-238 from the more explosive uranium-235 used in bombs. These analyses are so sensitive that scientists using the uranium/lead 'clock' have to worry about how much lead the meteorite may have picked up from the air in our cities, which is polluted by lead fumes from car exhausts!

Scientists at the Lunatic Asylum have probed carbonaceous meteorites with such finesse that they can say what happened *before* the birth of the solar system. And they've shown it is most likely that a supernova explosion did indeed start the collapse of a gas cloud to form our solar system, as outlined at the beginning of this chapter (although there are other forces which can make a cloud collapse to form a star and attendant planets, as mentioned on p. 137). A supernova spews out a small amount of newly-created atoms in its death-throes (Chapter 8), and some of the radioactive atoms live for only a short time – less than a million years. The Lunatic Asylum scientists have found indirect evidence that the carbonaceous meteorites did once contain such short-lived atoms. They conclude that a supernova must have exploded very close to the cloud that was to become our solar system, and less than a million years before the cloud started to collapse. Theorists had already calculated that a supernova would squeeze nearby gas clouds to the point of collapse within a million years of the explosion.

This evidence all fits the postulated chain of events too closely to be just coincidence. There's little doubt now that our solar system is the direct result of a nearby supernova explosion: we owe our existence to the spectacular demise of that anonymous, long-dead star.

The asteroids and their meteorite fragments are not the only reminders of our solar system's birth. Every few years, the ghostly figure of a comet will walk our night skies, striking terror into unsophisticated peoples for whom comets have always portended the deaths of princes: a comet appeared before Julius Caesar's assassination in 44 BC and another in AD 1066, as William of Normandy set sail to Hastings. But comets are more accurately a sign of birth, deep-frozen remnants from the early days of our planetary system.

Despite all its show, a comet is really a very insignificant object. All its activity stems from a small solid *nucleus*, a kind of large dirty snowball made of ices – not simply water ice, but also frozen 'ices' of ammonia, carbon dioxide and methane. Measuring a few kilometres across, it is very like the planetesimals which formed into the outer planets. And that's probably what it is. The comet nuclei have survived since those early days, way out beyond the planets where they were in no danger of becoming swept up by one of the forming worlds. Astronomers estimate that there must be something like 100,000 million comet nuclei out there, circling the Sun in huge orbits typically a thousand times farther out than Pluto. Despite the huge number of them, each comet nucleus is so small that all of them put together would scarcely make a planet the size of Jupiter.

Sometimes one of these comet nuclei is perturbed into a path that brings it in amongst the planets. As it comes within the orbit of Jupiter, the Sun's heat begins to melt the frozen ice. Water vapour and carbon dioxide boil away into the near-vacuum of space, to build up a huge glowing head around the nucleus. Although the head, or *coma*, contains very little matter, its gases spread out to a huge extent – sometimes to fill a volume larger than the Sun. The comet is now easily seen with a telescope, appearing as a fuzzy ball of opalescent light.

As it approaches the Sun, the comet brightens up. The outward force of sunlight and the solar wind push glowing gases and dust grains out of the comet's head, into a long 'tail'. This is the 'picture postcard' view of a comet: a glowing head sprouting a long luminous veil of a tail which stretches across the sky. Then the comet rounds the Sun, and heads back out into space – and because the tail is always pushed away from the Sun, a comet leaves the Sun travelling tail first!

The gravity of the planets – especially giant Jupiter – can, however, swing a comet round so that it is permanently trapped among them. Such a comet follows an elongated orbit that stretches out only as far as the orbit of Jupiter and Saturn; and its path brings it back to the Sun's vicinity every few years or decades. Its doom is now assured. On each close approach, the Sun will boil away about one-hundredth of its ices, and the nucleus will dwindle. Each appearance will be less spectacular; no longer will the comet be able to support the extravagance of a tail. Most known comets, incidentally, have reached this senile phase, and are dim, fuzzy tail-less creatures, visible only with a telescope. After a few thousand years, death will come. The nucleus either evaporates entirely, or the ices disappear to leave a small mass of dark rocky clinker.

The most famous of all comets is on its way to demise. It's named after the early eighteenth-century British astronomer Edmond Halley, who first realised that some comets return regularly to the Sun, but 'Halley's comet' had in fact been seen many times before. The first recorded sighting comes from the Chinese, who saw it in the winter of 1057 BC during a campaign of King Wu. At that time it was 300 times brighter than the brightest star, Sirius; but the comet has passed the Sun forty times since then, and it is now a much diminished object. There's been a lot of publicity for its next return in 1986, but unfortunately the played-out comet won't live up to many people's expectations.

Scientists are excited, though, because several spaceprobes will be sent to Halley's comet, to take the first close look at a comet. From Earth we can never see the actual tiny nucleus of a comet as anything more than a point of light. The probes should test our ideas on this, the most important, if least spectacular part of the comet. Leading comet expert Fred Whipple predicts the nucleus will appear as 'a smallish, potato-shaped object with a rough smudgy surface similar to the snow banks lining city streets in winter'.

Comet nuclei are just some of the many solid bodies found in space. The space between the planets is not simply empty or filled with gases; it's littered with solid debris of all sizes, from microscopic dust grains to asteroids. And the Earth is running into this junk all the time. Fortunately for us, the larger chunks are relatively rare.

The Earth sweeps up something like 100,000 tonnes of solid matter from space every year. Although this sounds a lot, it is distributed over a large globe – if this matter were spread out evenly over the Earth, it would make a layer only a hundred-millionth of a centimetre thick. Almost all of this matter comes down in an invisible rain of microscopically-small grains of dust; most of the remainder as rock particles about the size of grains of sand. The latter are heated up by friction with the air, and they burn away before reaching the ground. Their bright trails across the sky form what we call *meteors* or 'shooting stars'.

Comets jettison a lot of meteor particles into space as their ices evaporate. These particles spread around the whole orbit of the comet, and when the Earth crosses the orbit each year there is a 'shower' of meteors. If the Earth runs into a strong concentration of particles, the shower can be truly spectacular. In 1966, the Leonid meteor shower (consisting of particles from comet Tempel-Tuttle) astonished astronomers with a veritable storm: thirty meteors every second!

Larger chunks of rock can survive the fiery passage through Earth's atmosphere, and land more-or-less intact. As we've already seen, such *meteorites* are not the product of comets, but strays from the asteroid belt. And although they are invaluable to scientists probing the early history of the solar system, meteorites have a more sinister side. We don't often think about it, but here on Earth we live under a constant threat. At any time, a heavy stone could drop from the sky without warning: into our garden, on to our house – or onto our heads.

Thankfully there's no need to get alarmed. The Earth is so big that it's very unlikely that a meteorite will fall directly on a human being and there's no recorded case of a person being seriously injured by a falling meteorite. Meteorites have occasionally punched holes in buildings, however – part of the Bovedy meteorite lodged in the asbestos roof of a police station when it dropped on Northern Ireland in 1969.

A bigger meteorite than those just described could be lethal even if it didn't land directly on top of anyone, because it could explode like a bomb on impact. But this requires a really exceptional meteorite. The biggest stone that's ever been seen to fall landed in the Chinese province of Kirin in 1976. It weighed almost two tonnes. Hitting the ground only 30 metres (98 feet) from a group of commune members, it gouged out a crater 2 metres (6.5 feet) across, flung fragments of soil 150 metres (490 feet) and raised a mushroom cloud of dust and smoke. Yet no one was hurt.

It's only meteorites of a million tonnes or more that can cause real devastation – and they are so rare that it is unlikely one will fall anywhere in the world in our lifetimes. Once we are talking about objects so big it doesn't matter whether they are small asteroids or comet nuclei: both would pack the same punch. The world's most famous meteorite crater, Meteor Crater, Arizona, is a kilometre-wide hole that is believed to have been blasted out by a million-tonne iron meteorite some 25,000 years ago. As recently as 1908, a fragile comet nucleus of similar weight exploded in the air above the Stony Tunguska valley in Siberia. Although it dug no crater, the tremendous explosion flattened the forest trees for 30 kilometres (19 miles) around. Fortunately the region is only sparsely-inhabited, but one of the nearer farmers was thrown from his chair and knocked unconscious. He was 60 kilometres (37 miles) from the explosion.

Although no one can predict where or when the next big meteorite will hit the Earth, astronomers can calculate the average time we must wait between impacts. This doesn't mean they are regular – just as an average interval of ten minutes between buses means we may have to wait as little as two or as many as twenty minutes when we've just missed one.

A million-tonne asteroid or comet should hit Earth every thousand years or so, and for all the local devastation at Tunguska, such impacts really affect only a very small portion of the world. Larger meteorites could however affect the whole world. Controversial British astronomer Sir Fred Hoyle has recently proposed that a huge meteorite could throw up so much dust into the atmosphere that it would block off sunlight and cause an Ice Age on the Earth. Hoyle proposes that a new Ice Age could start at any time, if a large enough meteorite should land anywhere in the world. Very few scientists would agree with Hoyle's speculations, though. There's no evidence in favour of this theory, and the resurgence of glaciations (and the warmer interglacials between) has already been explained – very accurately – by the Milankovitch theory which relies on well-understood variations in the Earth's orbit and the direction of its poles (see Chapter 2).

But the very largest meteorites would cause widespread damage. And many scientists are coming to believe that it was a rogue asteroid which destroyed the dinosaurs, those giant reptiles which suddenly disappeared from the face of the Earth some 65 million years ago.

An asteroid 10 kilometres (6 miles) or more in diameter – the size of a city – should hit our planet every hundred million years on average. Forty to fifty such giant bodies must have smashed into our world since it was born. Such an impact must have affected the entire world – and in more recent times, had serious effects on the vulnerable living things which have come to populate the Earth. Geologists know of two really major extinctions of life on our planet. The more famous is the death of the dinosaurs. In itself, the extinction of these reptiles would not be so remarkable because the huge creatures must have been susceptible to even mild environmental changes, like a slightly cooler climate. But at this point, 65 million years ago, a whole range of other species was wiped out too, many of them sea-dwellers who should have been immune from most changes. The earlier tragedy, 225 million years ago, was even more drastic – although it's not widely known outside geological circles because it involved only the primitive life-forms around at the time.

These intervals – 65 and 225 million years ago – are certainly about right for us to invoke an asteroid as the killer. But this is unscientific speculation. No one – least of all the geologists – are going to be convinced that the dinosaurs were the victims of a cosmic catastrophe unless there is some stronger proof. But in 1979, a team of American geologists suddenly found themselves confronted by just such evidence.

Luis Alvarez and his team were investigating rocks from the Gubbio Valley in Italy, which were laid down on the sea-bed about 65 million years ago. A thin layer of clay separated the rocks laid down before the great extinction of life from those deposited just after. When Alvarez's team analysed this boundary layer, they were amazed to find it had a most unusual composition: in particular, it contained twenty-five times as much iridium as the other rock layers. Iridium is a relatively rare element on Earth – like its sister-element platinum – but it is much more common in meteorites. Alvarez tentatively concluded that this iridium must have come from space, and his results made the cause of the dinosaurs' extinction 'look more extraterrestrial than terrestrial'.

Geologists all around the world immediately began to inspect exposed rocks dating from the great extinctions, from Denmark to the United States to New Zealand. Everywhere, they've found this thin layer of clay, and everywhere the clay is rich in iridium. It's difficult to see how ordinary geological processes could produce such an iridium-rich layer, and spread it all around the world. Alvarez is now convinced that there's only one possible explanation: the dinosaurs must have been killed by the impact of a small asteroid.

Purely by chance, the Earth found itself on a collision course with an asteroid 65 million years ago. This huge rock, 10 kilometres (6 miles) in diameter, an interplanetary Mount Everest, ploughed through the Earth's atmosphere, briefly shining as brightly as the Sun in the sky. It hit the ground and exploded with the force of ten thousand hydrogen bombs. It gouged out a crater 200 kilometres (124 miles) across – a hole as wide as Iceland – blasting billions of tonnes of powdered rock into the atmosphere. This airborne dust floated for years, in a dense black pall blotting out all sunlight from the surface of the Earth. Eventually the dust settled, to form the tell-tale layer of clay, with its strange complement of extraterrestrial elements.

But the newly-cleared skies revealed a different, barren world. During the years of darkness, plants had died in their millions, deprived of life-giving sunlight. Animals suffered in turn. Herbivores starved to death as the supplies of their plant food dwindled; and the fierce carnivores which preyed on the herbivores ended up short of food, and perished too. Only a few species survived. Amongst them were small, warm-blooded creatures that were to evolve into the mammals.

Such a story of cosmic devastation reads very convincingly, and the evidence has convinced many scientists that this is how the dinosaurs died. But the evidence so far does not amount to conclusive proof. Geologists in particular are quite rightly demanding a lot more analysis and discussion.

They are naturally suspicious of anyone who wants to explain geological changes as a result of cosmic influences, because it's so often an easy way out: a neat, and almost untestable, way of explaining changes that are in fact caused by complicated geological events. A wayward asteroid is today's version of the unpredictable hand of God!

For all this, astronomers now know that our planet is not an absolutely secure haven. Every so often, a stray asteroid or comet nucleus must hit the Earth, and cause widespread damage. Even if the dinosaurs were not destroyed by an asteroid impact, the great chunks of rock in space must have caused devastation at some points in Earth's history. We are always in danger, though the chances of a collision in any person's lifetime are tiny.

If the birth of our solar system had been a tidier affair, we wouldn't be as exposed to such calamity from space – if the asteroids had coalesced into a planet between Mars and Jupiter, if the comets had accumulated into another, more distant Jupiter. But without the smaller debris which comes to Earth as meteorites, we would also know very little about one of the most intriguing mysteries of all: how our Sun and Earth were born.

— 7 —

Our Local Star

Deep in the hills of South Dakota lies the world's most sensitive solar telescope. But this is more than just a romantic description, for the telescope does lie literally *in* the hills, buried underneath 1500 metres (4900 feet) of rock down a working goldmine near the town of Lead. With this instrument – a 500,000 litre (100,000 gallon) tank of cleaning fluid – Raymond Davis of Brookhaven National Laboratory has been peering into the Sun's very heart for the past fifteen years. And his findings have made some astronomers uneasy about how well we understand our local star.

Davis's experiment sounds bizarre, but it's a vital check on our ideas as to how our Sun and other stars work. It exploits the basic difference between the Sun and its surrounding family: like all stars, the Sun has a powerful nuclear energy source, while its planets do not. The tank-telescope constantly monitors this generator at our Sun's core.

The difference between the Sun and its planets arose at the time of their birth, initially as a result of the young Sun's privileged position at the centre of things. So much of the collapsing gas and dust cloud ended up in the middle of the forming solar system that the gas became extremely compressed. Its temperature rose steeply. Eventually, it became so hot at the centre that nuclear fusion reactions became inevitable. The proto-sun turned into a star, generating energy from its own powerhouse and stopping any further collapse. But had it been less massive – about a fiftieth of its present mass, or twenty times the mass of Jupiter – the temperature could never have risen high enough to trigger nuclear reactions. Our Sun would have become a steadily-cooling 'failed star', forlornly circled by the frozen globes of its planets.

Today, we take the Sun's energy so much for granted that it's difficult to realise how much we depend on it. Not only does it shield us from the utter cold and darkness of space, but in less direct ways it has provided fuel and food throughout the entire period of Earth's history. Without this apparently limitless supply of energy, life would be an impossibility; but we hardly ever stop to question the Sun's constancy. What guarantees do we have that its energy will still keep coming, and at the same rate? We will be returning to the vexed question of the Sun's constancy (or otherwise) later in this chapter.

As with all stars, the key to our Sun's prodigious energy flow lies deep in its core. Compressed by the enormous weight of overlying layers, the gas at the Sun's centre is at a temperature of about 14 million °C (25 million °F). At these temperatures, atoms can't exist: the electrons are moving so quickly that they break away from their dominating protons altogether. The centre of the Sun – and the rest of it, for that matter – consists of a *plasma* of rapidly-moving atomic nuclei (protons and neutrons) and electrons.

Most of the nuclei are just single protons: hydrogen atoms stripped of their electrons. The high temperatures in the Sun's core cause these to collide repeatedly, and sometimes the collisions are so forceful that protons, assisted by the uncharged neutrons, are impelled to overcome their natural electromagnetic repulsion and bond together. The result, two protons and two neutrons, is a helium nucleus.

Luckily for our energy supply, this is a case where the whole is not equal to the sum of its parts. A helium nucleus is actually only 99.3 per cent as heavy as its four constituents, weighed individually. And so, every time four hydrogen nuclei fuse to make helium, a tiny bit of unnecessary mass is liberated. This mass is converted into pure energy in the heat of the reaction.

The mass-discrepancy involved in hydrogen fusion seems so small that we might expect only a trickle of energy to be generated by this means. But the Sun is so vast that it has been converting 4 million tonnes of its matter into energy every *second* of the past five thousand million years! Despite this mammoth rate of self-digestion, there is plenty of the Sun left. It has used up less than a few ten-thousandths of its mass by nuclear reactions.

It's hardly surprising that we on Earth would dearly love to tap nuclear fusion for our own energy needs. This highly efficient mass-into-energy conversion process is the key to how all stars work: in effect, they are all slow-running hydrogen bombs. The flood of energy gradually welling upwards from the core keeps the star shining and buoyed up against the potentially catastrophic inward pull of its own gravity.

But not quite all the energy wells up so slowly. Liberated in the reactions are chargeless and (probably) massless atomic particles called neutrinos, which travel away from the reaction at the speed of light. Because of their ghostly properties, they interact very little with matter: it's estimated that a neutrino could travel through light years of lead before being stopped. As far as the neutrinos are concerned the Sun is transparent. It takes them just two seconds to get from the Sun's core and away into space.

Astronomers can predict how many neutrinos should be produced by the reactions going on inside the Sun. If we could detect them, it would not only serve to tell us whether our ideas on the energy source of the Sun (and all the other stars too) are close to the mark; it would also double as a check on the Sun's reaction rate *at the moment*. In other words, neutrinos give us the chance to 'look' into the Sun's very centre.

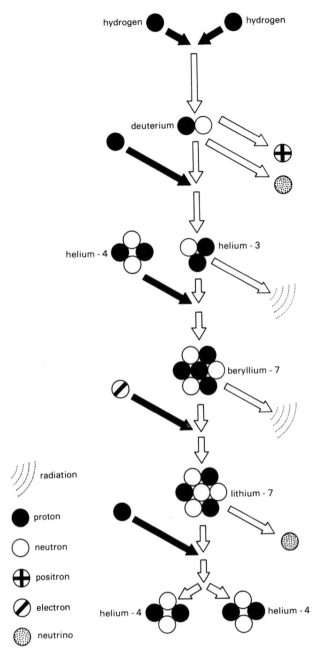

11 NUCLEAR FUSION AND NEUTRINO PRODUCTION *The Sun's energy comes from a chain of nuclear reactions which convert hydrogen nuclei (protons) into helium nuclei. During each stage of the main reaction, particles and energy are emitted. Only neutrinos can pass unaffected through the Sun's overlying layers, streaming out into space directly from the reaction at the speed of light. By capturing solar neutrinos and measuring their numbers and energies, astronomers hope to gain a precise picture of the conditions and reactions taking place in the Sun's core.*

This is just what Raymond Davis is doing with his tank of cleaning fluid (perchloroethylene) down the Homestake goldmine in South Dakota. Ten thousand million million neutrinos pass straight through the tank each second, but every so often one is stopped by a nuclear reaction as it collides with the nucleus of a chlorine atom in the tank. The result is to change the chlorine atom into an unstable, radioactive form of argon, readily detectable when the tank is flushed out with helium gas every few months. When Davis first devised his experiment, he made sure he shielded his tank well to prevent other fast-moving particles from getting in on the act – hence the overlying shield of 1500 metres (4900 feet) of rock. Then he waited to see if these elusive messengers from the Sun's core would turn up at the rate astronomers had predicted. Out of a hundred thousand million million million neutrinos passing every day through the tank, Davis expected that just one would produce an interaction.

After fifteen years, Davis's rates have proved disconcertingly low. Instead of trapping a neutrino a day, he finds the tank stops one every four days. The tank has been thoroughly tested out. The only easy explanation is that the Sun is producing only about a third of the number of neutrinos that it should.

This result is still worrying scientists quite a lot. If they can't predict how the Sun behaves, what chance do they have to really understand other stars? But as solar physicist John Parkinson wryly remarks: 'All astrophysicists have very vivid imaginations, and so there is no shortage of ideas for solving this neutrino problem. Some solutions can only be described as bizarre ...' We'll be looking at some of the less bizarre solutions later on in this chapter.

Davis's tank captures only the most energetic neutrinos coming from the Sun's core. To get a more balanced picture, astronomers would like to be able to pick up the lower energy neutrinos, too; but until recently, the ideal detector for these particles, the rare metal gallium, was in very short supply. However, as an important ingredient in the manufacture of semiconductors, solar cells and computer memories, it's now beginning to be produced in commercial quantities. But with a market price of £250,000 per tonne, it is still hardly cheap. Nevertheless, a team of astronomers has obtained 1.4 tonnes of gallium chloride to form the basis of the world's first gallium telescope. Appropriately enough, it will live next to Davis's original tank-telescope in the Homestake goldmine, and when fully assembled it should contain 50 tonnes of gallium – bringing the cost, after fifteen years of operation, to a staggering £3500 per neutrino captured!

It's a lot cheaper to ignore this hotline to the Sun's core and simply wait until the radiation produced by reactions slowly bubbles up. But it does take a very long time, and by the time it reaches the surface, the radiation has changed its character entirely. It starts off its long, tortuous trek as highly energetic X-rays and gamma rays. After travelling only a centimetre

(.4 inch) or so, each bundle of energy collides with a particle in the Sun's interior, bouncing off only to collide with another ... and another. The radiation pursues a zig-zag path upwards, losing energy by collisions all the time. On reaching the surface, 700,000 kilometres (435,000 miles) and a million years later, it will emerge not as powerful X-rays, but as low-energy light and heat.

The Sun's energy literally bubbles up at its surface – 'like boiling chip fat', says John Parkinson. Good photographs of the Sun's churning, gaseous *photosphere* (surface layer) show 'granulation cells' where currents rise and fall. In places, this rice-grain structure is interrupted by large, cooler regions where the bubbling appears hampered in some way. Because they're cooler than the Sun's 6000 °C (10,800 °F) surface, these *sunspots* look very dark – but at 4000 °C (7200 °F), they'd glow dull orange if we could see them on their own. By following the paths of sunspots across the disc, astronomers can see clearly how our local star spins. Instead of rotating as a solid body, the sunspots reveal that our gaseous Sun takes 33 days to spin once at the poles, but only 25 days at the equator. It's possible that this 'differential rotation' may be at the root of some of the Sun's periods of bad weather.

To see the Sun in more detail than this, the answer isn't to photograph it directly at all, but to spread out its light, wavelength-by-wavelength, with a spectroscope. Even when the Sun's light is dissected in this way, there's still an awful lot of it compared to the light we can study from other stars. This is just as well, for the extraordinarily detailed spectrum we get from the Sun provides a really firm basis for our studies of other, far more remote stars.

Taking a spectrum is just a sophisticated version of passing light through a glass or perspex prism to make a rainbow. A hot, glowing object – a tungsten lightbulb, for example – shows a rainbow spectrum. It reveals how light from the object is distributed across the different wavelengths (colours) which go to make up 'white light'. As we shall see in Chapter 8, the peak wavelength in a *continuous spectrum* like this is a good guide to the object's temperature. The Sun's continuous spectrum, produced by the hot, dense layers of gas just below its visible surface, peaks in the yellow region. This corresponds to a temperature of about 6000 °C.

But there's a far more sensitive temperature check than this. Unlike a lightbulb, our Sun's rainbow spectrum is crossed by thousands of thin dark lines, where light from the continuous bright spectrum has been absorbed. It's these *absorption lines* which really lay bare our Sun's secrets. They're giveaway signs that the hot layer of gases producing the continuous spectrum is wrapped in a thin, cooler envelope of far less dense gas: a region where the gases are free to maintain their own identity and reveal it through their own individual spectral signatures.

The spectrum of a rarefied gas is like a fingerprint. Instead of giving out (or absorbing) light right along the spectrum, each gas emits (or absorbs) radiation only at very sharply-defined wavelengths – a result of the particular atomic make-up of the gas. Since every gas is made differently, each has its own recognisable spectrum. In theory, a few characteristic spectral lines, not even the whole spread, should be sufficient to identify a gas: but in the Sun's case, it's not quite that easy. The Sun's gases comprise almost all the known varieties of chemical element, and although some spectral signatures are obvious, disentangling the weaker ones is a bit like trying to identify 90 different tunes being played simultaneously on 90 separate pianos!

Because there is so much available light from the Sun, astronomers can

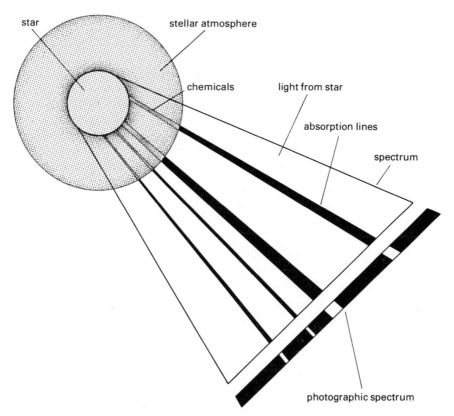

12 HOW THE SPECTRUM OF THE SUN/A STAR IS PRODUCED *Sunlight or starlight dispersed into a spectrum (by a prism or diffraction grating) shows a characteristic rainbow band crossed by dark, narrow absorption lines. The 'rainbow' comes from hot, dense gas deep down in the star's atmosphere, which emits radiation over a whole range of wavelengths. Absorption lines arise from cooler, more rarefied gases in the upper layers of the star. The atomic structure of these overlying gases permits them to absorb from the deeper levels only at certain specific wavelengths. This registers as an absence of light – an absorption line – against the rainbow continuum. Each kind of atom produces its own characteristic pattern of lines, allowing astronomers to identify which different gases are present in a star's atmosphere.*

disperse its spectrum very widely by diffraction gratings (generally used instead of prisms) and study it in exceptionally fine detail. The curtain-fold appearance of crowded parts of the spectrum is deceptive. It looks as if the Sun is overflowing with gases such as iron, sulphur, magnesium and calcium, whereas these substances are actually very rare. About three-quarters of the Sun is hydrogen, almost all the rest is helium: all the other elements together make up only two per cent. It's just that the Sun's surface is at the right temperature for these rarer gases to produce strong spectral lines. In the same way, the spectra of stars very much hotter or cooler than the Sun can look very different, even though roughly the same mix of elements is present in every case.

The absorption lines aren't just a guide to the temperature of a star. A careful investigation of spectral lines, and particularly those of our nearby Sun, can tell astronomers just what it's like to be on the surface of a star. As well as revealing the mix of elements, and the proportions of each, the lines also indicate the physical conditions on the surface: temperature, pressure, density, gravity, speed of convection ('bubbling'), and even whether there are strong electric or magnetic fields in the area.

Take the spectrum of a sunspot, and the power of spectroscopy is revealed to the full. Not only do spectra of these dark regions reveal their low temperatures and inhibited patterns of flow, but they show up the culprit too. Controlling the whole area of a sunspot is an intensely powerful magnetic field (see below). In some cases, the field can be many thousands of times stronger than that of the Earth.

Sunspots are the most obvious sign that our local star suffers change. The number of spots on the Sun ebbs and flows, building up to a maximum approximately every eleven years and dying away in between. (More accurately, the total cycle takes twenty-two years, because the Sun's general magnetic field reverses direction every eleven years.) The first spots of each cycle appear at high latitudes and gradually work down towards the equator as the cycle continues. The biggest spots of all grow to enormous sizes – many hundreds of thousands of kilometres across – and when the Sun is dimmed by mist, they can be seen without any optical aid at all. The Chinese astronomers were recording naked-eye sunspots as long ago as 28 BC, and accorded them such descriptions as 'the shape of a three-legged crow' or 'as large as a melon'. They noted that some lasted for several months, which is indeed the case for the largest sunspots of all.

The real value of sunspots lies in their role as tracers of the Sun's changeable magnetic weather patterns. They're the most evident sign that a great loop of magnetic field, like a horseshoe magnet, has pushed up through the Sun's surface layers and is arching upwards into space. Where the loop cuts the surface, it cools and slows down the churning gases. Above, in the Sun's lower atmosphere, great loops of incandescent gas, *prominences*, dance down the field lines as if in a giant fireworks display.

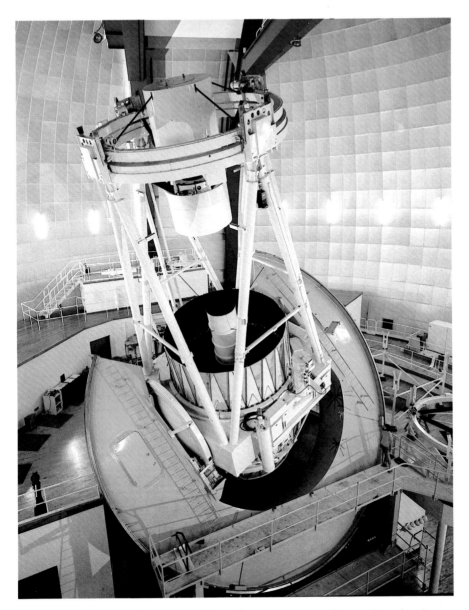

1 The Anglo-Australian Telescope, sited on Siding Spring Mountain near Coonabarabran in New South Wales, is the world's fourth largest optical telescope. Its size can be judged by the man standing near its large 'horseshoe yoke' mounting, which steers the telescope to point to any part of the sky. The 3.9 metre (153 inch) diameter mirror at the base of the framework tube weighs seventeen tonnes, and reflects light to a focus 12.7 metres (41.6 feet) up the tube. Completed in 1974, the Anglo-Australian Telescope was the first of the modern generation of computer-controlled optical telescopes; the astronomer does not work directly on the telescope in the cold night air, but sits in front of a console in an adjacent warm room.

2 ABOVE LEFT: *The Very Large Array of radio telescopes near Socorro, New Mexico, mimics a single radio telescope 25 kilometres (15 miles) across. Its 27 individual radio dishes – each 25 metres (82 feet) in diameter – are arranged along three straight lines forming a huge Y-shape across the desert. They are connected by buried electronic waveguides to a computer which combines their results to produce radio maps of the sky showing finer details than even the largest optical telescopes can 'see'.*

3 BELOW LEFT: *Earth is the 'Blue Planet' of the solar system, its colour due to its unique oceans of liquid water. The blue expanses of the South Atlantic and Indian Oceans dominate this beautiful view, photographed by the Apollo 17 astronauts on Man's final trip to the Moon in December 1972. Between the two oceans lies the familiar outline of Africa, stretching up to the Mediterranean Sea (top) and the Arabian Peninsula (top right). Wisps of white cloud veil much of the southern hemisphere, while the dazzling-white ice sheets of Antarctica (bottom) are enjoying the continuous daylight of a polar summer.*

4 BELOW: *Apollo 11 view of the Moon's far-side is dominated by the 80 kilometre (50 mile) wide crater Daedelus. The slumping and terracing of its walls reveal that this is an ancient crater, formed by the impact of a meteorite some 4000 million years ago.*

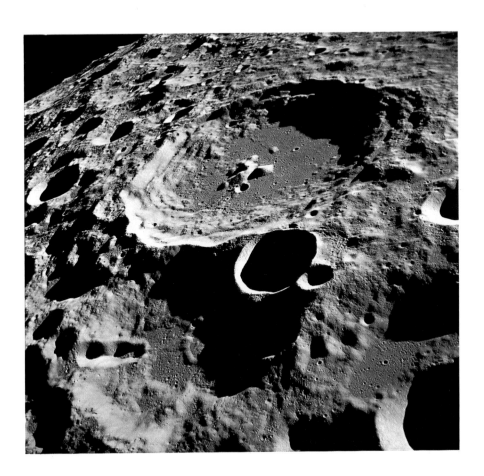

5 BELOW: *David Scott, commander of the Apollo 15 mission to the Moon, salutes the US flag after a successful touchdown on 30 July 1971. Behind him towers the 4000 metre (13,000 foot) high Mount Hadley, which lies 5 kilometres (3 miles) away across the plain.*

6 ABOVE RIGHT: *Our neighbour planet Venus is perpetually shrouded in thick clouds of sulphuric acid droplets, here photographed from a range of 65,000 kilometres (40,400 miles) by the American Pioneer Venus Orbiter which became an artificial 'moon' of Venus in December 1978. Series of photographs show that the clouds are racing around the planet from east to west (right to left) at a speed of 350 kilometres (220 miles) per hour.*

7 BELOW RIGHT: *The hidden surface of Venus beneath its clouds was revealed by a radar set aboard the Pioneer Venus Orbiter: its data were used to construct this contour map. Two-thirds of the planet is covered by 'rolling plains', broken here and there by a few lower-lying basins and by two large plateaux named Ishtar Terra and Aphrodite Terra. The twin peaks of Beta Regio are probably active volcanoes. Most of Venus's topographical features are named after famous women in history and mythology.*

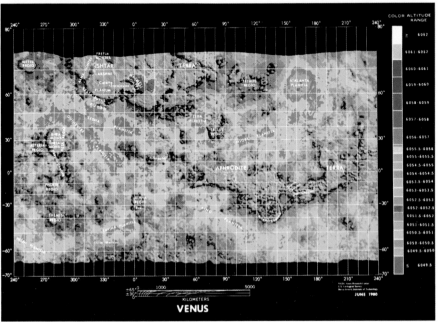

VENUS

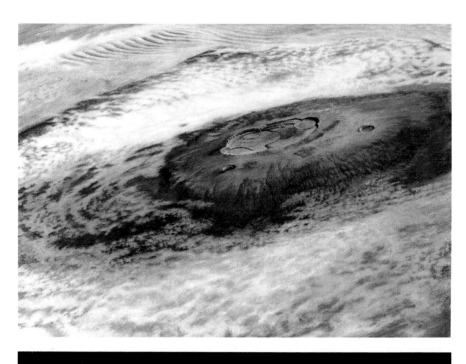

8 ABOVE LEFT: *The martian volcano Olympus Mons is far larger than any volcano on Earth, although Mars is only half the size of our planet. Olympus Mons is 25,000 metres (82,000 feet) high, and measures some 400 kilometres (250 miles) across its base. This oblique view is a colour-tinted version of a photograph sent back by the American Viking Orbiter which reached Mars in 1976.*

9 BELOW LEFT: *Jupiter's swirling cloud patterns and its Great Red Spot form the backdrop for two of its moons in this view of the giant planet photographed by the Voyager 1 spaceprobe in February 1979 from a range of 28 million kilometres (17 million miles). The orange satellite seen against Jupiter's disc is Io; to the right is the brilliant ice-covered Europa.*

10 ABOVE: *An eruption from the volcano Loki on Jupiter's moon Io ejects sulphurous gases 100 kilometres (62 miles) above Io's surface. Voyager 1 took this close-up photograph from a range of only 500,000 kilometres (310,500 miles) on 4 March 1979, shortly after its first pictures of Io had surprised scientists by showing the unpredicted volcanic plumes.*

11 *Saturn, its rings and its satellites made a spectacular sight as Voyager 2 approached the planet in August 1981. Weather patterns, less distinct than Jupiter's, are faintly visible on Saturn's globe, while the rings are resolved into a multitude of narrow ringlets. The satellite Rhea (above left) is Saturn's second-largest moon; Dione (centre left) and Tethys (top left) are almost twins, each about 1100 kilometres (680 miles) across. The shadow of Tethys is falling on Saturn's globe as a small dark 'spot'.*

If two arching magnetic loops of opposite polarities should meet, the result is a colossal explosion in which the tremendous magnetic energy is liberated – a *solar flare*.

Sunspots, prominences and flares all follow the 11-year (or 22-year) solar cycle. What causes the cycle isn't certain: magnetic fields are somewhere near the bottom of it, but they're probably not the whole story. Astronomers used to believe that the Sun's differential rotation steadily 'wrapped up' the weak magnetic field in its surface layers, coiling it into ropes which grew in strength until they 'bobbed up' through the surface itself. As the winding process continued, the ropes slowly migrated towards the Sun's equator until the 'north polarity' ropes of one hemisphere neutralised the opposite 'south polarity' ropes, making it possible for the cycle to begin again. Today's picture is not so clear-cut: some astronomers claim that the driving force behind the magnetic field may instead be jet streams, fast flows of gas at particular latitudes.

What is much more certain is that the Sun's weather has an effect on us. In Chapter 2 we saw how solar activity gave rise to beautiful displays of aurorae when its streams of fast-moving charged particles were channelled down onto our poles. More sinisterly, we found how cruelly buffeted the Earth could become when, during a magnetic reversal, it temporarily loses its own magnetic field. Our magnetic field undoubtedly protects us from some of the worst of the Sun's weather, but it can't keep it all out. When flares erupt on the Sun, they each give out as much energy as ten million hydrogen bombs. There is no way in which Earth can avoid the fast-moving particles produced in the blast, which reach us a couple of days later and crash down through our upper atmosphere. By affecting the ionosphere (the level in our atmosphere off which we 'bounce' our radio transmissions) flares can severely disrupt radio communications and have been known to cause complete power blackouts. While this is serious enough for aviation and communications links in general, a flare's effect on space-bound humans is potentially still more damaging. The intense burst of X-rays produced in a flare – enough to bump up the Sun's X-ray luminosity by a factor of 100 in seconds – reaches Earth in minutes, and could cause serious space sickness, even death, to a crew who were travelling out of our protective atmosphere at the time. Luckily, the Sun is continuously monitored for imminent flares, and there's an efficient worldwide 'flare warning' network.

It's becoming clear that the Sun's weather must affect us in more subtle ways, too; even to the extent of causing changes in our own notoriously unpredictable weather. Although scientists are loath to admit correlations in the absence of an efficient mechanism to explain them, it really does seem that Earth's weather is better at sunspot maximum. This is backed up by records of exceptionally high cricket scores (only possible during periods of very fine weather) and the years associated with outstanding wine vintages.

Scientists intrigued by the Sun's influence on us have been busily blowing the dust off old records to check if there are any historical correlations. It's a controversial area, because the reason for the interaction, if any, isn't obvious. Princeton professor of physics Robert Dicke speaks for many scientists when he says: '... there have been very many claims of the influence of solar activity on the climate. Most of these claims are highly questionable and the whole subject of solar-terrestrial relations has developed the aura of a science of ill repute.'

One coincidence which is difficult to explain away is the observed lack of sunspots, and indeed of any other solar activity, during the period 1645–1715. People were actually looking for sunspots and recording the numbers of aurorae over the whole seventy-year period, and so the low numbers can't be the result of indifference. This 'Maunder Minimum' (named after the famous Victorian astronomer E.W. Maunder, who first noticed it) coincided with a remarkably cold period in Northern Europe. During the 'mini ice-age', it became so cold that the River Thames actually froze on several occasions, allowing delighted Londoners to participate in boisterous 'frost fairs' on its icy surface.

Leading solar astronomer Jack Eddy believes that the Sun is more usually in this 'quiet' state. He bases his claim on measurements of the amount of radioactive carbon (carbon-14) found in annual growth rings in trees of known date. Its concentration should be greater when the Sun is quiet, because these are periods during which the Sun's magnetic field cannot protect us from the cosmic rays from beyond the solar system, which produce the carbon-14. During the Maunder Minimum, the carbon-14 count was indeed exceptionally high. If the explanation is correct, then tree-rings tell us that the Sun has been in a generally depressed state for most of the past 3000 years. The past two hundred years has seen an unusual burst of activity – and as a result, we should be enjoying a climate that's better than normal!

Lest this be too depressing, there's equally good evidence that in the very long term, at least in its sunspot cycle, the Sun is very reliable. Measurements of the thickness of mud deposits ('varves') laid down in an Australian lake 700 million years ago seem to indicate that the eleven-year cycle was alive and well even then. Not only that, but it was very strong – and it evidently had enough effect on the weather to dictate the thickness of sediments deposited.

During the past 1980 solar maximum, the Sun has been observed as never before. For the entire year, an international team of scientists set up home in 'Experimenters' Operations Facility' at the Goddard Space Flight Centre, Maryland, to monitor data coming in from the specially-designed Solar Maximum Mission satellite which orbits Earth at a height of 570 kilometres (360 miles). The scientists were particularly interested in the most violent manifestations of solar activity, solar flares, whose energetic X- and gamma ray radiation never reaches the ground. Glancing up at the serenely-shining Sun after three months of sharing the 'Experimenters' Facility' with five other teams of scientists, plus minicomputers, disc drives, tape units, printers and colour graphics display terminals, solar physicist Chris Rapley remarked: 'It's difficult to relate that thing up there to what we're doing.'

Nevertheless, there's undoubtedly a very close, unseen relationship between the Sun and the Earth, whose extent we are just beginning to gauge. And it's hardly surprising that there are such strong ties, for our planet lies embedded within the outermost parts of the Sun's atmosphere.

Nowhere are the Sun's faint, outer extensions more spectacularly revealed than in a total solar eclipse. Unlike lunar eclipses, which occur whenever the Moon moves into Earth's shadow – and are consequently seen over wide areas of the world – eclipses of the Sun are rare and awesome events. It's a peculiar coincidence which makes them possible: the fact that the Sun and Moon appear very nearly the same size in the sky. Of course, the Sun is very much bigger (400 times, to be exact) but then, it's also 400 times further away. And so, when the Moon passes *exactly* between the Sun and the Earth, it can blot out the Sun's brilliant disc completely. But if it's to work, this alignment must be very precise. The Moon's orbit is slightly angled to the Earth's path around the Sun, which reduces chances at the outset. Then the Moon mustn't be too far from the Earth: if it is at the far point of its elongated orbit, its disc will appear too small to blot out the whole Sun (and we have an 'annular' eclipse). And because the Moon *is* so near, changing your position on the Earth's globe can make a great difference to the perceived alignment.

Although total eclipses happen several times a year, in any one place they are very rare. The last to be seen in the United Kingdom was in 1927, and this was visible only to northern England. The next will be seen on 11 August 1999 – but only from the southernmost tip of Devon and Cornwall. Nor do they last long. The entire period of totality, when the Moon exactly overlaps the Sun, may be as short as a few seconds, and the absolute maximum is only seven minutes.

It's difficult to get an idea of the terrors that such eclipses held for our ancestors, until you have seen one for yourself. Even some cynical astronomers today have confessed to feelings of 'primitive fright' when the Sun's light is suddenly swallowed up and brilliant day is turned into eerie half-night. The spectacle must have been infinitely worse for our ancestors, so much more at the whim of Nature than we are today. Had the Sun gone away forever? What had eaten it? Would it eat them too? Even today, a few modern races still believe that an eclipse is caused by a dog gobbling up the Sun. And there are some Indian cultures who believe pregnant women must stay indoors for the entire duration of an eclipse (throughout its 'partial' phases too when the Moon covers only part of the Sun, and wherever in the world it happens to be) for fear that it will do harm to their developing child.

The glory of an eclipse, the sight that generates fear, awe, wonder, panic or simply a feeling of absolute beauty in the minds of its watchers, is the breathtakingly sudden appearance of our Sun's pearly outer atmosphere around its blacked-out disc. The *corona* is exquisite. You can trace its pearly tendrils for four or five sun-widths across the sky, until it melts into the grey half-light. In the inner corona, vivid pink prominences surround the eclipsed Sun like a herd of ant-eaters. While you watch, all around is deathly quiet. Birds and animals have gone to sleep; plants and flowers curl up for the night. The wind has dropped; the air is cold and still. But just as suddenly as it began, it's over: what seems to have lasted only seconds has actually been going on for minutes. There's one more spectacle in store – the 'diamond ring effect' as the Sun's brilliant surface flashes out from behind a tiny hollow in the Moon's edge – and then it is daytime once more.

It's not hard to understand why people travel thousands of kilometres to see an eclipse. As a 'wonder of Nature' it is unrivalled; and it's never a case of 'once you've seen one, you've seen them all'. There are dozens of different factors which make every eclipse different. However, it comes as something as a surprise to learn that, even today, teams of astronomers still trek to out-of-the-way eclipses in the time-honoured tradition. Surprising, because a great many observations of the corona can be done from space: the tenuous gas is extremely hot (millions of degrees) and it emits large quantities of X-rays. But eclipses still give optical and infrared astronomers their very best chance of getting to grips with the innermost workings of the corona. And their results are impressive, despite the unreliability of the necessarily portable equipment and the degree of improvisation involved. As American astronomer Jay Pasachoff reflects: 'Astronomers have to be very versatile at an eclipse site.... It helps to be a combination carpenter, mason, machinist, electronics engineer, optical physicist, and more.'

The corona is the origin of the Sun's local environment, the solar wind. Although it's a gentle breeze compared to the howling gales around some stars, the solar wind rips through space at a very respectable 450 kilometres per second (a million miles per hour!). It's this incessant onslaught which keeps comet tails pointing away from the Sun, and buffets the planets with storm-force gusts whenever there's a solar flare.

But the solar wind is a protective shield as well as an unrelenting force. This became clear in July 1981, as Pioneer 10, our farthest-travelled spaceprobe, crossed the 'Silver AU' (astronomical unit) mark, a point twenty-five times more distant from the Sun than Earth. Pioneer 10's cameras (the first to photograph Jupiter in 1973), have long since shut down, but many of its other instruments are still working and transmitting back data. Scientists at NASA's Ames Research Centre in California were surprised to find that even at the 'Silver AU' mark (halfway between the orbits of Uranus and Neptune) the solar wind is still blowing strong, with little or no decrease in speed, despite predictions that it ought to die away just outside the orbit of Jupiter. They now forecast that it should stretch to a hundred times the Earth–Sun distance, well beyond the orbit of Pluto, before merging with the interstellar gas. In fact, it looks as if the whole solar system is enclosed within a gigantic 'magnetic bubble' generated by our Sun, which helps to keep out some of the more damaging radiations from the rest of our Galaxy. In the event of a nearby cosmic catastrophe, such as a supernova outburst (see Chapter 8), this invisible bubble could help save our lives.

The extent of the solar wind means that all the planets are likely to share a similar environment; a reassuring fact for space-mission planners of the future. 'If you were living on any other planet', maintains pioneering space-scientist James Van Allen, 'you would find a solar environment surrounding it like that surrounding Earth. This would include: a constant solar wind, buffets by solar magnetic storms, and in many cases, radiation belts.' And so when Pioneer 10 passes the orbit of Pluto in October 1986, it should notice little difference in its surroundings. But early next century, it will leave the safety of the bubble for the open waters of space. As it speeds away, watching our solar system dwindle into nothingness, it might detect a strange effect predicted by John Simpson of Chicago University. 'Like a giant cosmic lung', he forecasts, 'this bubble may breathe, expanding and contracting with each eleven-year solar cycle.'

For all its far-reaching effects, the eleven-year cycle appears to be no more than a recurrent and superficial skin-ailment on the Sun's surface. But there are indications of deeper change. As we already know (see Chapter 2), the Sun has gradually warmed up by some fifty per cent over the period of the Earth's existence, an effect which Earth has successfully counteracted by steadily altering the composition of its atmosphere. We have every chance, at this rate, of holding our own until 5000 million years in the future when our dying Sun will swell rapidly to become a red giant (see Chapter 8). Faced with sudden change like this, the atmosphere will be unable to cope, but it is unlikely that we will still be around to suffer the consequences.

The Sun's gradual heating-up is no surprise to astronomers, because it's an expected part of the life-cycle of all stars. More disconcerting are changes or departures from the norm that haven't been predicted. And one of these brings us right back to where we started: the case of the missing neutrinos.

As we recall from the beginning of this chapter, the nuclear reactions at the Sun's heart are producing only a third of the neutrinos they ought. At first rather shaken by this discovery, astronomers were quick to offer explanations. Perhaps the Sun was more 'mixed-up' at its centre; maybe the core was clamped in the vice-like grip of a strong magnetic field, or was it spinning very rapidly? Could the core's chemical composition be slightly different from that expected? All of these scenarios produce fewer neutrinos, but it was the most 'exotic' suggestion which swept the boards at the time. A very easy way to decrease the neutrino flux is to lower the nuclear reaction rate. A lower reaction rate implies a cooler, quieter core. So the neutrinos we detect, winging their way to us directly from the Sun's heart, tell of a comparatively sedate state of affairs. On the other hand, the sunlight we receive – which has taken a million years to penetrate from core to surface – was evidently produced when the Sun's core was far more active. In other words, the Sun's very core is variable.

If this explanation is correct, at least it won't be those of us living now who will have to suffer the repercussions of the Sun's inconstancy. We are reaping the benefits of radiation produced long ago. It will instead be our descendants who'll feel the effects of the current 'moribund' state of our Sun's core. But with the uncertainty in predicting just how the Sun affects the Earth, it's hard to know what these will turn out to be.

Some support for these dismal predictions has come from astronomers who watch the Sun 'shiver'. Every object has a natural resonant frequency at which it vibrates, and the Sun is no exception. The 'solar oscillations' must be driven by some kind of 'hammer', and the Sun's core is the most likely culprit. But the trouble with the solar oscillations is that they take far too long: the Sun's surface wobbles back and forth by 4 kilometres (2.5 miles) every 2 hours and 40 minutes, nearly twice the expected period. The only way to reconcile these slow shivers with theories of our local star is to reduce its current core temperature by some ten per cent.

One leading expert on the neutrino problem, however, describes all these explanations as 'cocktail party solutions'. Princeton University's John Bahcall reckons that they are all contrived simply to dispose of the neutrino problem without much consideration for anything else. But this can't be said for Jack Eddy's suggestion. He believes that the Sun is shrinking.

Eddy, of the Harvard-Smithsonian Centre for Astrophysics, is one of the world's leading solar astronomers. He and a colleague, Aram Boornazian, have ploughed through over a hundred years of solar timekeeping records made with transit telescopes at the US Naval Observatory and the Royal Greenwich Observatory. Sifting through meticulous measurements made every day at noon by teams of trained observers, Eddy and Boornazian have concluded that the Sun is shrinking at a rate of about 2 arc seconds – roughly one-thousandth of its diameter – every century.

Such a rapid rate of contraction can only be temporary. 'It's unrealistic to assume this will continue', cautioned Eddy. 'It does seem to imply that the Sun is oscillating in some way. However, going farther back into time to find an expansion will be difficult since the records get dimmer and dimmer.'

Eddy's discovery would, if substantiated, solve the neutrino problem at a stroke. Some of the Sun's energy would come from gravitational contraction, so that the demands on the nuclear furnace at the core would be lessened. 'I won't be surprised to see a future model [for energy production] where the Sun is getting its energy from several mechanisms', Eddy forecasts. So, if the core is just one energy source of many, it needn't be as active as we have assumed. Its temperature could be lower. And we wouldn't expect to see many neutrinos.

These claims have, quite naturally, excited a great deal of controversy. Other astronomers have gone through the same data that Eddy used, taking into account additional factors, and haven't found the effect. Some have looked to earlier eclipse records to check if there has been any gradual variation in the Sun's size, only to find there's been no change. Other groups still, using data similar to Eddy, have reported a very much slower rate of shrinkage in the Sun's diameter. Most astronomers, it is fair to say, view any claims for a shrinking Sun with a sceptical eye.

Nevertheless, the last ten years have shown us that we don't know our local star quite as well as we thought. The old Sun still appears to have surprises up its sleeve. Some astronomers find this worrying. Perhaps our knowledge of all stars could be shaky as a result? But let a solar physicist have the last word: 'So, what's wrong with the Sun? Are the observations wrong or the theory incorrect? We don't know. It's as simple as that. All we can do is to carry on working and hope to improve both.' With an approach like John Parkinson's, we can't go too far wrong.

— 8 —

The Secret Life of Stars

Ask anyone to name a star pattern, and it's odds-on that even the most unscientific person will come up with at least one – their 'birth sign' constellation. If pressed, many people will go on to explain that they are a 'typical Taurus, Gemini, etc'; and that the positions of the Sun, Moon and planets against the remote backdrop of stars at the time they were born had a powerful hand in moulding their characters. Without a doubt, belief in astrology is on the increase. Despite – or perhaps, because of – this impersonal microchip age, many people are turning to it and to other 'mystic' paths to help give themselves perspective and direction. But do the stars really influence us? Modern science has answers that are even stranger than astrology.

It is easy to see how astrology developed such strong roots in our subconscious. The early astronomers – Babylonians and Chaldeans living over 4000 years ago – mapped the stars by joining them, dot-to-dot fashion, into an imaginative array of constellation patterns. Some they named after objects of similar shape; others took on the names of legendary heroes and heroines. With a few later additions, we still use the same constellation patterns to this day.

The patterns formed the dial of a huge celestial clock which wheeled from east to west every night. And the clock doubled as a calendar, because different constellations became visible as the year unfolded, and familiar patterns became identified with spring, summer, autumn and winter. Today we know that this repeating parade of stars troops past us as we tread our yearly path around the Sun; but it must have seemed remarkable to those not 'in the know' that the planting season, the growing season and the harvest were watched over by their own particular stars.

More impressive, the ancient astronomers could tie in other natural events with the appearance in the sky of certain stars and constellations. Egyptian astronomers, for instance, noted that the autumn appearance of the bright star Sirius heralded the onset of the essential Nile floods. To the uninitiated, this must have seemed miraculous, and it was but a small step to assert that Sirius's appearance somehow *caused* the floods. And so astrology was born.

Early astrology concerned itself only with politics and governments. But if the stars influenced the destiny of a country, what of its people? 'Personal' astrology evolved to cover the fate of individuals, and this is the form in which we usually encounter it today. Indeed, today's astrologers defend their subject vigorously, claiming that it is exceedingly ancient, extremely complicated, and has millions of believers. They forget that the same was once said of alchemy. But then, what of the scientific tests astrology has been put through? None of these has shown anything of significance, except a widely-reported analysis performed by Michel Gauquelin which gave some support to the idea that planetary influences can shape our character. Yet other researchers, using similar data, have found no such connection.

If the constellation patterns were real, if the planets and stars lay at the same distance, if there were really a physical connection between them, if, if . . . then astrology just might have a claim to a scientific basis. But the Sun and planets are our near neighbours; the stars lie billions of kilometres away. Constellation patterns are made up of unconnected stars which just happen to lie in roughly the same direction far off in space, usually separated from one another by hundreds of light years. To say a planet is 'in' a constellation is rather like watching a fly crawling across a window-pane against a distant vista of mountains, and to then assert that the fly is 'in' the mountains.

Leading astronomy educator Jay Pasachoff sums up the astronomers' point of view by saying: 'Astrology is meaningless, unnecessary and impossible to explain if we accept the broad set of physical laws we've conceived over the years to explain what happens on the Earth and in the sky. Let's all learn from the stars, but let's learn the truth.'

As always, truth is stranger than fiction. The stars *do* have an influence on us: an influence more profound than any guessed at by the astrologers. But to find out the nature of this influence, we literally have to look inside the stars themselves. We need to discern the overall picture.

This would seem to be wishful thinking when we consider the distances to the stars. The nearest is a dim southern hemisphere star called Proxima Centauri, which lies at a distance of 4.3 light years. In astronomers' verbal shorthand, this rolls off the tongue quite easily: but in terms of everyday distances, it becomes utterly staggering. A 'long' car journey – covering a distance of 500 kilometres (300 miles) – lasts five hours. Travelling at the same speed to Proxima Centauri would take 48 million *years*. Even our fastest spacecraft would be hard-pressed to make the journey in much under 80,000 years. And Proxima Centauri is a good deal closer than most stars.

Then there is the immense range in star-types, which makes it all the more difficult to discern the picture. You can get a feel for this by looking carefully at the sky on a really clear, black night, and noticing how the stars differ in their colour and in their brightness. Take brightness first. The Greek astronomer Hipparchus systematised this 2000 years ago; he grouped the stars into brightness classes 1 to 6, with the most brilliant stars of first magnitude, and the faintest visible to the naked eye of magnitude 6. This system was quantified in the last century, so that each magnitude is 2.512 times fainter than the one before it. A difference of five magnitudes means that one star is exactly a hundred times fainter than another. As a result, the magnitude system had to be extended to *negative* magnitudes – the Sun, Moon, most planets and a handful of the brightest stars are all technically brighter than zero. The system has stood up well, although it has been greatly extended at the faint end as larger telescopes have captured dimmer and dimmer stars. From the faintest stars the Space Telescope will detect (magnitude *plus* 29) to the dazzling brilliance of our Sun (magnitude *minus* 26.5) there lies a brightness span of over 10,000 million million million times!

Not all of this enormous range is real. The spread in distances amongst the stars accounts for a lot of it, but even when distance is taken into account, we find that the stars still have a very wide span in their luminosity. Our Sun's luminosity, not surprisingly, turns out to be very average. It is far outshone by a few rare celestial searchlights over 100,000 times brighter, and far outnumbered by millions of stellar glowworms only one ten-thousandth as bright.

The spread in the luminosity of stars is paralleled by their spread in temperature – or colour. Colour acts as a crude star-thermometer in the same way that a lump of metal in a furnace betrays its temperature when heated, first glowing red-hot, amber, gold, and finally white. When you spot reddish Betelgeuse or Aldebaran in the winter sky, you are looking at two 'cool' stars, with surface temperatures of 'only' 3000 °C (5400 °F). Orange Arcturus is lukewarm at 5000 °C (9000 °F), while our Sun and golden Capella are a warm 6000 °C (10,800 °F). White stars, like dazzling Sirius, are hot (10,000 °C, 18,000 °F), but the hottest stars of all (20,000 °C and above) have, like Rigel, a bluish tinge.

Use a telescope to ensnare more stars and, as well as confirming these ranges in temperature and brightness, we find even more differences. Some stars live alone; others live in clusters or swarms. Some are single, while others are double or multiple. There are stars which change slowly in brightness. Others even explode.

If we are really to make sense of this variety amongst the stars, we need a way of looking still more closely at them. But how? Even in today's monster telescopes, they appear as no more than pinpricks of light. But telescopes are the bluntest part of a modern astronomer's weaponry, merely flux-buckets to capture the maximum possible amount of light. Once an astronomer has entrapped the elusive lightwaves, his most important task is to wring them dry of the information they hold. And one of the most tried, tested and efficient ways to do this is to spread out the starlight into a spectrum.

Just as spectroscopy revealed the secrets of our local star, so it has laid bare some of the mysteries of our Sun's distant neighbours. A star's spectrum is broadly similar to the Sun's, a bright band crossed by dark lines; but like a fingerprint, each star has a unique spectrum. Analysis of the message encoded in a star's spectrum can tell accurately of its temperature, its size, its chemical make-up, its speed through space, and any peculiar traits it might harbour, such as sharing gas streams with a close companion star. And when all this spectral harvest is gathered in, astronomers find even more bewildering variety amongst the stars. There are celestial heavy-weights, some fifty times – and one perhaps more than 2000 times – heavier than our Sun. Some tiny stars weigh only a twentieth of the Sun's mass. Then there are stars which, if placed in our solar system, would engulf all the planets out to the asteroid belt. These red giant stars are the only exception to the rule that stars appear as 'pinpricks through the telescope': a cunning new technique called speckle interferometry can just reveal the bloated and star-spotted surfaces of a few of the closer ones. At the other end of the scale, we find crushed, superdense dwarf stars only the size of planets.

Double, single, giant, dwarf, variable, exploding, massive, superlumi-nous, dim, searingly-hot, cool, multiple ... where is the overall picture we set out to find? Faced with a problem like this, scientists categorise and classify to see what broad patterns emerge. Although, like botany, astro-nomy's heyday of classification lies deep in the last century, it was through these beginnings that we now know why stars are the way they are.

The early spectroscopists pigeonholed each star they observed with two identification tags: luminosity and temperature. The latter let them group the stars into a range of temperature-dependent 'spectral classes'. Logically enough, the classes started with 'A' and worked through the alphabet – our Sun, for example, is a G star – but later refinements showed that some classes were superfluous, and others wrongly-ordered. The list now reads (working from hot to cool) O B A F G K M. It's hardly surprising that various dubious mnemonics have been devised to help people to remember the sequence, but only 'Oh, be a fine girl (or guy), kiss me ...' seems to have stood the test of time.

Working independently, Ejnar Hertzsprung in Denmark and Henry Nor-ris Russell in the United States then compared the luminosities and spectral types of hundreds of stars on a simple graph. Instead of scattering all over the page, the stars formed an orderly sequence, confining themselves to a tight band which crossed the graph from top left to bottom right. A few stars lay above and below, but the conclusion was clear. There was a relationship between the luminosity of a star and its temperature. The Hertzsprung-Russell Diagram had shown up a real pattern where confusion had once reigned.

Their work was certainly a triumph, and illustrates well the fact that

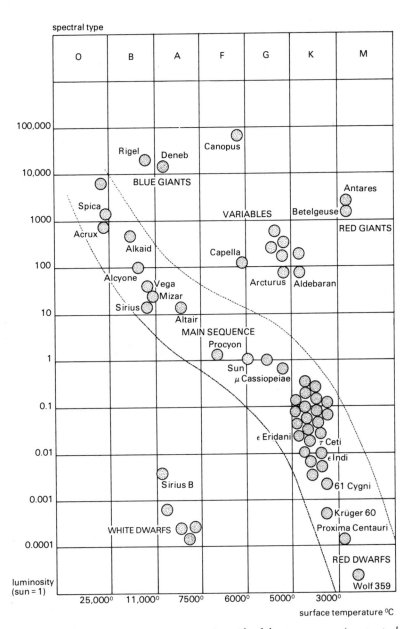

spectral type

| | O | B | A | F | G | K | M |

100,000

Canopus

Rigel · Deneb

10,000

BLUE GIANTS

Antares

Spica

VARIABLES

Betelgeuse

1000

RED GIANTS

Acrux

Alkaid

Capella

100

Alcyone

Vega

Arcturus Aldebaran

Mizar

Sirius

10

Altair

MAIN SEQUENCE

Procyon

1

Sun

μ Cassiopeiae

0.1

ε Eridani

τ Ceti

0.01

ε Indi

Sirius B

61 Cygni

0.001

Krüger 60

WHITE DWARFS

Proxima Centauri

0.0001

RED DWARFS

luminosity
(sun = 1)

Wolf 359

25,000° 11,000° 7500°   6000° 5000° 3000°

surface temperature °C

13 THE HERTZSPRUNG–RUSSELL DIAGRAM *A graph of the temperature (or spectral type) of hundreds of stars against their luminosity reveals an interesting pattern. Instead of scattering all over the diagram, most stars fall into a well-defined band of temperatures and brightnesses, running from top left to bottom right. All stars on this* main sequence band *– including our Sun, which lies almost centrally – are undergoing the same physical processes. Stars which lie off the main sequence, like the red giants (top right) and the white dwarfs (bottom left), generate their energy in a slightly different manner, and so do not fit the usual temperature-luminosity relationship. This graph, which reveals a great deal about the evolution of stars, is called the Hertzsprung–Russell Diagram in honour of its two independent discoverers.*

scientists classify objects not for the convenience of lumping them into forgotten piles, but in order to highlight barely-evident physical truths. The particular truth which Hertzsprung and Russell had illuminated was actually a story: the story of how stars are born, live and die.

The variety amongst stars is mirrored by that found amongst people. An alien with little time to spare, hovering in his flying saucer above Trafalgar Square, would see a tremendous range of human beings – babies, adults, youths, toddlers, the elderly – and he could be forgiven for being confused. But if he realised that all these different forms were stages in the life of every human being, then in a very short time he could construct a 'scenario' to describe how humans evolve from cradle to grave. So it is with the stars. Although their lifetimes are so long compared with our own that we can seldom hope to see a star actually changing with time, we can at least attempt to piece together the different stages of its life, and predict how it will behave at each stage.

Today's astronomers manage these predictions with the aid of high-speed computers: huge 'number-crunchers'. To the average astronomer, a star is not a remote and tantalising orb, but a straightforward lump of gas, subject to the same laws of physics as those governing the behaviour of a lump of gas in the laboratory. How are stars born? Astronomers like William Kaufmann III are typically unromantic: 'Using all the laws of physics, a great deal of mathematics, and a computer to speed up the calculations, we simply ask what happens as a ball of gas contracts under the influence of gravity. Astrophysicists think they are on the right track with their ideas about the birth of stars because the computer's answers agree with observations.'

Aesthetically though, the spectacle of star formation is infinitely more glorious than the sight of a shrinking gas-ball. We are surrounded by sites of star-birth: glowing gas clouds like the Orion Nebula, the Rosette Nebula, the Trifid Nebula, the Carina Nebula . . . their swirling draperies are visions that no astronomer can ever tire of. But even these glowing 'star-cradles' are a relatively late stage in the story of the birth of a star.

The saga begins between the already-existing stars, where space is pervaded by a cold, incredibly tenuous gas composed mainly of hydrogen atoms. Its average density is so low that a spoon-sized scoopful would contain only one atom. But time, and the inexorable pull of gravity – from stars and from the gas atoms themselves – have conspired to mix these atoms with the tiny soot grains between the stars and heap them up into colossal clouds of gas and dust (see Chapter 11). Deep inside these dark clouds, many light years down, the conditions are right to make stars. It is pitch-black; it is still; it is cold; there is nothing to stop the gas collapsing under the pull of its own gravitation – except the very nature of the gas itself. Astronomers have never pretended that star formation is an easy matter, and several of today's 'theoretical starmakers' are frankly astonished that stars manage to form at all. If you compress a ball of gas, it doesn't straightforwardly collapse under gravity, but responds rather like a squeezed balloon by bulging somewhere else. Manchester University astronomer J.E. Dyson muses '... so many are the effects opposing star formation that the difficulties seem almost insuperable. However, it's obvious from the structure of the Galaxy that nature has no such difficulties!'

Astronomers generally agree that dark clouds would sit there forever unless given an initial kick to make them shrink. One way to kick them is to explode a supernova nearby: the shock of the blast is enough to make a gas cloud cave in. Another way is by slowing them down as they travel around the Galaxy. Slowing-down and speeding-up is something which seems to be built-in to the galactic traffic regulations, as we shall see in Chapter 11.

Once a gas cloud has started to collapse, there is very little anyone can do to stop it. Collapse is swift, particularly towards the centre where the gas is densest. Here it is also shielded from disruptive radiations by the dark outside cloak of dust particles, and the gas cloud's interior breaks up into denser fragments. These, in turn, break up over and over again. In the end, there may be hundreds of fragments, all contracting independently and growing steadily hotter as they shrink. Each is the embryo of a star.

Until recently, protostars lived only in computers, because although astronomers had every reason to believe that they existed in reality, no telescope could penetrate the secretive mantles of the dark clouds. But now we have the means to pry. Astronomers can now pick up long-wavelength infrared and radio waves which can slice through the smog; and they find, nestling inside the clouds, groups of warm, dense, fledgling stars, cocooned in skeins of dust. In 1980, radioastronomer Patrick Palmer of Chicago University went so far as to claim: 'We believe we have isolated a protostellar object for the first time.' Scanning the gas-rich Orion region with the world's largest steerable radio telescope near Bonn, West Germany, Palmer's team detected a warm cloud only twenty times the size of our solar system. They found it was contracting in size at a rate of 5 kilometres (3 miles) per second (18,000 kilometres, 11,000 miles per hour). 'Every time a new, more powerful radio telescope comes along, someone finds a cloud smaller than has been seen before and calls it a protostar,' says Palmer. 'Most of those are far too large to form a single star or even a double star. But this cloud's small size, temperature and motions make me think it will almost certainly collapse to form a star.' But Palmer isn't expecting instant results. 'Of course, that could take tens of thousands of years.'

A protostar continues to shrink under the relentless inpull of its own gravity, getting steadily hotter towards the centre as layer upon layer of its gas is compressed. All around, sooty dust grains jostle in a swirling disc, sometimes sticking together to form the seeds around which its planets will grow. But even now, it is not a star. Not until the central temperature has reached 10 million °C (18 million °F) does the situation dramatically change. At this point, the dense gas at the core can no longer avoid the sudden onset of nuclear fusion reactions. A wave of energy surges swiftly outwards through the protostar to balance the gravitational inpull, and the millennia of collapse come to an end. The lump of gas is now stable, and it has its own energy source to keep it that way. In short, it is a star.

'Stable' is only a relative term when applied to young stars. Like youngsters of any variety, they are restless in their behaviour, varying suddenly and spectacularly in their brightness as they rearrange their structure or abruptly hide behind the shifting clouds of dust which still envelope them. As time passes, their powerful radiation drives away these sooty wrappings, and the young stars emerge to the gaze of the Universe packed in a tight cluster. Surrounding them are the tattered outer layers of the original gas cloud, which once afforded them protection as they formed. Now, excited by the strong ultraviolet light from its young progeny, it glows resplendently as a nebula. But its glory is destined to be shortlived. The fate of this leftover gas is to be propelled by the glowing energy of its offspring into the vastness of space, where it will perhaps form into stars the next time round.

Wreathed in youth-revealing shreds of dusty gas, the stars in the cluster cling close together under their mutual gravity for their first few million years. There are dozens of such 'open clusters' in our skies. Most beautiful of all is the Pleiades. These are the 'Seven Sisters', for six or seven are visible even to the unaided eye, but the cluster contains in all some 250 stars, born only 60 million years ago. Others, like Praesepe – Latin for 'The Beehive', but known in China as the 'Exhalation of piled-up corpses' – are considerably more ancient, perhaps 1000 million years old.

Given enough time, most open clusters will disperse as their members break the bonds of one another's gravity and make their own way through space. Because of the way in which stars are born, our Sun must once have belonged to such a cluster; today, we cannot even recognise any of its nestmates. But not all stars make such a clean break. The majority, in fact, are formed so close to another star that they remain together as a binary pair, or even a multiple group, for life. The new technique of speckle interferometry, which involves photographically 'freezing' Earth's shifting air-currents by very short exposures, can show up very fine details – such as the discs of nearby giant stars, as mentioned earlier. But perhaps its most important achievement is that this technique has revealed that many hitherto 'single' stars are actually very close doubles. Unless the two components of the pair are so close that they are practically touching, though, the presence of a companion has little effect on a star's subsequent life.

By far the greater part of this life is uneventful. The star shines because of the energy liberated by the hydrogen-into-helium fusion reactions going on in its core; energy sufficient to perfectly balance the star's gravitational forces which, unopposed, would make it collapse completely. 'Seen in this context' says infrared astronomer David Allen, 'a star is a subtle invention of nature to avoid that embarrassing singularity in the laws of physics we call a black hole.'

Nevertheless, not all stars generate energy at the same rate, and nor do they live for the same length of time. The key to the difference lies in their range of masses. Stars much heavier than the Sun have higher core temperatures; they rip through their nuclear fuel at a profligate pace, and burn themselves out correspondingly quickly. Our Sun, and stars less massive than it, are cooler-hearted and more conservative: they can look forward to lifetimes lasting thousands of millions of years instead of the frenetically-lived few million years of their weightier cousins.

Knowing this, we can now regard the Hertzsprung-Russell diagram as a kind of stellar 'weight-chart', as well as an outline life-story. Once a star is born, its mass fixes its temperature and luminosity – and so its position on the graph. Massive stars are hot and bright; less massive stars are cool and dim. That's why stars fall on a tight, diagonal band – the 'Main Sequence' – going from top left to bottom right on the H–R diagram: it is a descending scale of mass. The tightness of the line tells us that all stars on the Main Sequence are undergoing the same physical processes, although at different rates.

There are, however, a few stars whose temperature and luminosity set them apart from the Main Sequence. This immediately tells a scientist that they must work in a different way from 'normal' stars like the Sun. Their small numbers also demonstrate that these 'abnormal' stars last only a short time. In fact, this 'abnormality' is just a passing phase every star has to go through, particularly as it approaches the end of its life. The most prominent dissenters are the very bright, but cool *red giant* stars, which lie at the top right of the H–R diagram; and the hot but dim *white dwarfs* at the lower left.

Even a star's dying moments are dictated by its mass. Low-mass stars like the Sun run out of fuel after some ten thousand million years. At this time, their central hydrogen reserves – one-tenth of the total mass – have been converted to helium. But only in these central regions are temperatures high enough for the fusion reactions to occur. The star has lost its energy source, and gravity reasserts itself, crushing tight the dead helium core. These internal traumas have little effect on a star's appearance until the core's temperature has risen so steeply under compression that a thin shell of hydrogen bordering it suddenly starts burning to helium. The star now has a short-lived power source, sited much higher up from the centre. There is less outer gas in the way to impede the outward flow of energy, and the effect is dramatic. Engulfing its nearest planets, the star billows out under the surge of unleashed energy, finally settling down in a new state of balance a hundred times bigger than it was before. This fate will one day befall our Sun – and its family. Earth may be spared, but Mercury and Venus will almost certainly be gobbled up and incinerated as our star becomes a red giant.

Some red giants are so huge that a ray of light would take almost an hour to cross them. Their cool surfaces are covered with starspots (similar to sunspots) as wide as Earth's orbit around the Sun, and their matter streams away into space in a galeforce 'stellar wind'. But for all their vastness, they're as insubstantial as thistledown, as rarefied as the best laboratory 'vacuum'. With most of their matter concentrated in the collapsed core, gravity has little hold on their outer layers. They wobble like cosmic blancmanges as their surface layers slowly pulsate in and out, causing them to change uncertainly in both size and brightness. Orion's brilliant Betelgeuse – 'the armpit of the Sacred One' – is one such long-period variable star which brightens and fades by a factor of two over a period of about a year.

But it seems that there's a limit to what an old star can take. In the end, it very gently puffs off its distended atmosphere. For an instant of cosmic time, a few tens of thousands of years, we see the red giant's remains as a planetary nebula – an ever-expanding shell of gas excited to shine by the now-revealed core. Even that soon disperses into space. All that remains is the core: a fiercely-hot, ultradense cadaver of a star called a white dwarf, whose fortunes we will follow in the next chapter.

If low-mass stars go out with a whimper, then massive stars go out with a bang – literally. After its spendthrift years burning hydrogen to helium on the Main Sequence, a massive star finds itself bereft of fuel in exactly the same position as its lighter cousins; and in the same way, it expands to become a red giant. But here's where the difference comes in: a massive star can muster together far greater gravitational powers. These it brings to bear on its inert core, squeezing it ever more tightly and causing its temperature to climb from tens of millions °C to a searing 100 million °C (180 million °F). At this point, the core springs back to life. It's now hot enough for another fusion reaction to take over, and the helium 'ash' is compressed into carbon, producing a main source of energy. The star behaves almost 'normally' once more, but its highly extravagant fuel consumption soon leaves it without a power source again, and on the way to becoming a red giant once more. Yet again, gravity tries its hand. This time, it succeeds in crushing the carbon nuclei making up the core into the heavier elements oxygen and neon, and the star gains a temporary power source for the third time. And so it goes on. The core grows steadily smaller, hotter and denser, while the inside of the star comes to resemble the layers of an onion, with concentric shells of different, newly-created elements. The star's luminosity and temperature during this unsettled period place it well above the Main Sequence on the H–R diagram. Here, we do find many short-period variable stars, the Cepheid variables, whose regular pulsations show that they are indeed expanded, massive stars.

A massive star continues living on borrowed time until the central part of its core is made up of iron. Then it tries the fusion trick for the last time: but it doesn't work. Nuclear reactors here on Earth tell us that iron fusion takes *in* energy rather than giving it out; and the star, determined to get the reaction started, attempts to take in energy the only way it knows – by collapsing still further. The effect is pandemonium. As the core's temperature rises to 5000 million °C (9000 million °F), the iron nuclei smash against one another and disintegrate, undoing in seconds what the star has taken millions of years to achieve. Unsupported, the outer layers start to cave in, only to be hurled into space at 12,000 kilometres (7400 miles) per second by the sudden heating and by the raging torrent of neutrinos produced in the breakdown. For a matter of weeks, a supernova like this will glare thousands of millions of times brighter than an ordinary star, even outshining its own galaxy, before taking months to fade into oblivion.

Not all supernovae are produced in this way, but astronomers are moderately confident that so-called 'Type II' supernovae are the death-throes of such shortlived massive stars. The problem is that supernovae are rare, and most of our information comes from those observed in distant galaxies. The last to be seen in our own Milky Way was 'Kepler's supernova' of 1604, and shortly before that, 'Tycho's supernova' of 1572 – a star-death we must be grateful for, because it jettisoned the great Tycho Brahe into transforming astronomy from an ancient to a modern science.

Before then, supernovae were widely believed to be portents of doom and destruction. Writing about the supernova observed by Arab astronomers in the year AD 1006, Ali ibn Ridwan records: 'I will now describe a spectacle which I saw at the beginning of my studies. This spectacle occurred in the zodiacal sign Scorpio, in opposition to the Sun ... Because the zodiacal sign Scorpio is a bad omen for the Islamic religion, they bitterly fought each other in great wars and many of their great countries were destroyed ... At the time when the spectacle appeared, calamity and destruction occurred which lasted for many years afterwards ...'

And so it appears that, via the life and death of stars, we have come back full circle to astrology. But it's an appropriate point to consider once again the influences which the heavens have on us, for, of all the stars, the role played by supernovae is the most fundamental. We've already seen (Chapter 6) that the powerful shockwaves generated by a nearby supernova some 5000 million years ago probably triggered the formation of the Sun and its planetary system. More than that, we owe our very existence to the funeral pyres of massive stars; and the same may hold true for every lifeform in the Universe.

The reason lies in the unique ability of supernovae to work as chemical processing plants. The early Universe was made up only of the two simplest elements, hydrogen and helium; these were the sole constituents of the first stars – as we can tell from the spectra of the oldest stars in our Galaxy. In those far-off days, there can have been no planets, and no life, for the complex elements required for building these structures were not yet present.

But inside the most massive of these early stars, the seeds of growth were being sown. In a desperate bid for survival before the fuel finally ran out, these stars steadily synthesised heavier elements – carbon, oxygen, neon – in their cores; elements destined to be spewed out into space when the first generation of stars met their violent deaths as supernovae. From the exploded crucible, a rich mix of heavy elements – with nuclei up to the weight of iron – gushed out into the swirling gasclouds earmarked for the next generation of stars. And in the heat and shock of the supernova holocaust itself, still heavier elements came into being: substances like gold, lead and uranium whose very rarity here on Earth and elsewhere in the Universe testifies to the extreme circumstances under which they were born.

The second generation of stars were enriched with the ashes of their ancestors, and they, too, exploded as supernovae, adding to the pool of 'building-block' elements. And so it went on, until stars no longer formed alone. In time, enough atoms of iron and oxygen and silicon and carbon clumped together to form the kernels of the first planets surrounding these stars. At last, there were secluded corners of the Universe capable, in some cases, of sheltering that fragile thing called life.

Life had to wait until there was sufficient carbon. Of all the elements, only carbon is capable of forming and sustaining the complex chemical bonds which support life. And most astronomers agree that the stable environment of a planet is needed to give life the space and time required to grow, to develop, and to explore.

This time, we really have come full circle. We *do* have links with the stars in a way which is far more basic, far more intimate, than any astrologer would dare claim. We now know that our bodies are made up of pure star-stuff. To enable us to live, millions of stars have had to die.

— 9 —

Star Corpses

On 28 November 1967, a young researcher at Cambridge stared in amazement at the paper chart emerging from her radio telescope. Signals from the sky, amplified by the telescope, were kicking the recording pen across the paper at precisely separated intervals, each just over a second in length. Jocelyn Bell had discovered a new kind of object in the Universe – the first *pulsar*.

But these first pulses from the regularly-ticking radio 'clock' in the sky didn't immediately excite attention. The discovery of pulsars illustrates that modern astronomy is a partnership between the researcher and his complex and sensitive equipment; and most 'discoveries' soon turn out to be faults in the equipment. Astronomers must check with the utmost care before publishing sensational new results.

December 1967 saw the Cambridge radio astronomy observatory seething with activity, as researchers tried to pin down the pulses to an equipment fault or to interference. Team leader Tony Hewish, who had conceived the new sensitive radio telescope in 1965, even interrupted his Christmas lunch to make critical measurements of the pulses. He says 'there was always a lurking disbelief that the pulses were real.'

Over that busy Christmas period, though, Hewish and his team were forced to conclude that the pulses did come from a cosmic source, in the constellation Vulpecula. The exact pulse timings changed slightly over the weeks, in exactly the way that the Earth's motion should have affected signals coming from beyond the solar system. And Jocelyn Bell, searching through the paper chart from the telescope (a chart now literally kilometres in total length) found three more pulsing sources. For all their unexpectedness, pulsars were evidently a fairly common kind of astronomical object.

As team leader, Tony Hewish was awarded the Nobel Prize for the pulsar discovery, sharing it with fellow Cambridge radio astronomer Sir Martin Ryle. In many ways it was more a human than a technological achievement. Many astronomers would have fed their telescope output straight into a computer and would have rejected unexpected radio 'noise' out of hand as interference. Hewish insisted that the output be recorded on a chart and be inspected by eye, since the human eye, as opposed to a computer, is good at recognising patterns. Jocelyn Bell's painstaking involvement meant that she spotted a small burst of 'scruff' on the chart soon after the telescope began to operate. Many astronomers, being as human as the rest of us, would have forgotten about the scruff – it wasn't what the telescope had been set up to find. But Bell and Hewish let their curiosity lead them on. They fitted up a faster chart recorder to record more detail in November, and found that the 'scruff' consisted of those unbelievably regular pulses.

Jocelyn Bell in fact relates a cautionary tale, showing how easy it is in astronomy, as in everyday life, to miss the big chance by dismissing an odd result as a mistake. A radio astronomer 'at another observatory' had been surveying the sky several years before when the pen of his chart recorder started dashing regularly across the paper every second. He muttered 'oscillation in the equipment', and in the time-honoured way of fixing electronic apparatus, he kicked it. The pen stopped. The astronomer was pleased; he put on his coat and went home. Only years later did he realise that he had been the first person on Earth to detect a pulsar. The honours went to the painstaking team of Bell and Hewish.

The nature of pulsars became clearer a few months later, when David Staelin and Edward Reifenstein in the United States discovered radio waves from a pulsar at the heart of a twisted gas cloud called the Crab Nebula. Astronomers have long known that the Crab Nebula is the exploded remains of a supernova which Chinese astronomers recorded as a temporary 'guest star' in the year AD 1054. The Crab Pulsar was evidently the core of this old star, left in the centre as the star's outer layers exploded around it.

Thirty-five years earlier, American astronomers Walter Baade and Fritz Zwicky had proposed the daring theory that a supernova's core might end up as a compact star made up entirely of neutrons. They are thousands of times smaller than atoms, and so a *neutron star* would be correspondingly much more compact, and much denser, than a normal star. A neutron star weighing as much as the Sun would measure only 25 kilometres (15 miles) across, its matter so condensed that a piece of neutron star matter only the size of a grain of salt would weigh 100,000 tonnes – as much as a fully-laden supertanker.

The discovery of the Crab Pulsar confirmed that pulsars are indeed neutron stars, and it confirmed theorist Tommy Gold's ideas on why neutron stars should appear to pulse. Gold proposed that the radio pulses were not caused by the star itself pulsing in and out, but by its spinning. The star beams out radio waves in two opposed directions, and as it spins, the beams flash across the Earth, making the source appear to 'pulse' regularly in exactly the same way that a lighthouse lantern appears to flash with monotonous regularity as it swings its beams of light around. The pulsar's radio waves come from electrons which 'broadcast' as they move

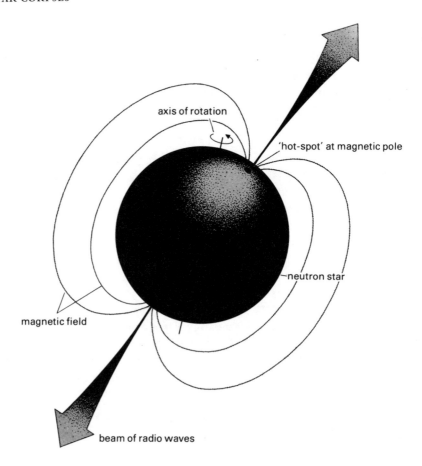

axis of rotation

'hot-spot' at magnetic pole

neutron star

magnetic field

beam of radio waves

14 RADIO EMISSION FROM A SPINNING PULSAR *A spinning neutron star* (pulsar)
*produces regular pulses of radiation in much the same way that a rotating lighthouse
lantern appears to flash. The neutron star has a strong magnetic field which is not aligned
with its axis of rotation. Near the two magnetic poles, fast-moving electrons generate
radio waves which travels outwards in two narrow beams. As the star rotates, the beams
swivel around, and when one of them sweeps past the Earth we pick up a 'pulse' of radio
waves. The star's continuous rotation produces a very regular sequence of pulses.*

through the neutron star's magnetic field – a field that astronomers calculate
to be a million million times stronger than the Earth's.

Spinning objects keep rotating at exactly the same rate, unless some
outside force affects them. Gold's theory thus explains the uncanny precision of the pulsar's pulses, which keep time better than a modern quartz
wrist watch. But precision timing over the years shows that all pulsars are
very gradually slowing down. Their magnetic fields reaching out into space
are like supple rubber bands: anchored to the neutron star at one end, and
to the stationary interstellar gas at the other, the field gradually brakes the
tiny star's rotation.

The Crab Pulsar is the youngest known. It spins thirty times every second. And it is also slowing down more quickly than any other. Its rotational energy is converted to magnetic energy and to speeding up electrons to very high velocities. These electrons circulate in the huge magnetic aura around the pulsar, and produce the light of the Crab Nebula. The Crab Pulsar powers the nebula in much the same way that a spinning electricity generator powers a fluorescent light.

Occasionally, though, the Crab Pulsar abruptly 'spins up'. The neutron star suffers a star-quake, contracts slightly, and as a result temporarily increases its spin rate. Such *glitches* allow astronomers to test their ideas of neutron star interiors, in the same way that earthquakes tell of the internal structure of the Earth. The bulk of the star is made of neutrons in a 'liquid' state, contained within a thin solid crust, and surrounded by an atmosphere of iron vapour only a few centimetres thick. Gravitation is so strong at the neutron star's surface that mountains on the solid crust cannot stand more than a few millimetres high; and it would take more energy than a man produces in his lifetime to climb a 'hill' a millimetre high!

Another pulsar which suffers glitches lies in the constellation Vela. Surrounded by the expanding gas clouds of a supernova which exploded some 10,000 years ago, the Vela Pulsar is the second-youngest known. Both the Crab and Vela pulsars are young enough to shine in ordinary light as well as radio waves: the Crab Pulsar can be seen on ordinary photographs, while the older Vela Pulsar is right at the limit of even large telescopes. It was first detected in 1977 with the help of electronic light amplification equipment attached to the Anglo-Australian Telescope.

A few pulsars are in orbit about other stars. The most interesting lies in the constellation Aquila, and is so close to its companion that they whirl around one another in only $7\frac{3}{4}$ hours. By measuring the changing length of its pulses as the pulsar pursues its orbit, American astronomer Joe Taylor has determined the orbital period with phenomenal precision – to an accuracy of a ten thousandth of a second in $7\frac{3}{4}$ hours!

Since the discovery of this binary pulsar in 1974, its orbital period has very slightly decreased, meaning that the pulsar and its companion (thought to be a radio-quiet neutron star) are gradually drawing closer together. There is only one possible reason. Albert Einstein had predicted in 1915 that two orbiting objects must always lose energy, radiated away as 'ripples' in their gravitational field. This gravitational radiation has a negligible effect in the solar system, and in ordinary double star systems. But the strong gravitation in the binary pulsar pair should make the system lose energy fast enough to alter the orbit significantly – and the predicted change is exactly the amount that Taylor measured. Appropriately enough, he announced the result in 1979, the centenary of Einstein's birth.

Radio astronomers estimate that there are around a million radio-emitting pulsars in our Galaxy. That means that one must be born every ten years, on average. But supernovae occur much more rarely than this, exploding in our Galaxy every thirty to fifty years, according to the best estimates. Either the number of pulsars has been overestimated, or possibly pulsars can be produced in other circumstances, without the fury of a supernova explosion.

Pulsars do not continue their radio transmission for ever. After a pulsar has slowed down to rotate in a couple of seconds, its radio emission begins to falter. Of the 320 pulsars that have been detected, most have periods between one and three seconds; the slowest known radio pulsar has a period of just over four seconds. By the time a neutron star is 10 million years old, it is no longer a pulsar, but merely a small dark, undetectable lump in interstellar space.

Only the heaviest stars end their lives as neutron stars. A lightweight star like the Sun does not explode as a supernova. It gently puffs off its outer layers to form the beautiful shell of a planetary nebula, while its core settles down to a small inactive corpse. Such a *white dwarf star* is not nearly as compressed and bizarre as a neutron star, though it is strange enough in everyday terms. A white dwarf consists of two-thirds the Sun's mass compressed into a globe the size of the Earth. A sugar cube of matter from it would weigh a tonne.

White dwarfs do not have nuclear reactions going on at their centres, but they do shine brightly enough to be seen in telescopes. When first born, they have the incredibly high temperatures of a star's centre. Although white dwarfs cool down very rapidly from these multi-million degree temperatures, their cooling rate slows down progressively as their temperature drops. As a result, we see many whose surfaces are still hotter than the Sun's, glowing at white-hot temperatures of 10,000 to 100,000 °C (18,000 to 180,000 °F). As they cool further, their colours gradually change through yellow, orange and red, but astronomers refer to them all as 'white dwarfs' for convenience.

The matter within a white dwarf is very unusual. Its density shows that matter is squeezed more tightly than ordinary atoms can be packed: in fact, the electrons are stripped from the atomic nuclei, and the electrons are packed together as tightly as possible. It's the electrons that give the star strength to hold itself up against its own gravitational inpull, while the atomic nuclei are simply scattered through this electron 'sea' like currants in a fruit cake. Because of the odd properties of such *degenerate* electron matter, a heavier white dwarf is actually smaller than a lightweight. And, as Indian-American astrophysicist Subrahmanyan Chandrasekhar first realised in 1931, it also means that there is an upper weight limit to white dwarf stars. Any white dwarf heavier than the Chandrasekhar limit of 1.44 Suns would collapse completely. We now know that an overweight white dwarf would collapse to become a neutron star, unless it were even heavier than three Suns, in which case its collapse would continue indefinitely, and the star would shrink to become a black hole (see Chapter 10).

Many white dwarfs are in orbit about ordinary stars. The best-known is the companion to Sirius – and because this star is the 'Dog Star', the white dwarf Sirius B is sometimes called the 'Pup'. Friedrich Bessel first announced that Sirius has a faint companion back in 1844; he couldn't see it, but discovered that Sirius was being perturbed by the gravitation of another body of about the Sun's weight orbiting it every fifty years. American telescope maker Alvan Clark discovered Sirius B accidentally in 1862, when he was testing a new high-quality telescope made by his father's firm. Astronomers found it hard to believe in such small, dense stars, though, until the structure of matter was unravelled by physicists in the 1920s. Now that neutron stars and black holes are 'respectable', white dwarfs seem very unremarkable members of the Galaxy.

In the 1960s, however, astronomers realised that white dwarfs are responsible for one of the most exciting of sky sights: a nova explosion. 'New stars' (*novae* in Latin) have always puzzled and intrigued men, and have left their mark in history. The appearance of a nova challenges our instinctive feeling that the slowly-wheeling heavens are permanent; they have even been taken as portents.

The greatest of the ancient Greek astronomers, Hipparchus, was inspired by seeing a nova in 134 BC to make the first star catalogue, so that he could check on other stars' disappearance or reappearance. As well as noting their positions, he recorded their brightnesses, inaugurating the system of star magnitudes that astronomers still use today (see page 96).

The most influential nova may have been one which flared up in 5 BC. This star may well have been the Star of Bethlehem. Certainly Jesus Christ was born several years BC; and according to St Matthew's commentary, the Magi saw a star which guided them to Jerusalem, and then to Bethlehem. The Magi were 'wise men', and in Mesopotamia their wisdom would have included astrology. If Matthew's account is true, rather than devised to fulfil existing prophecies, there are only two possible 'Stars'. One was the very close approach of the planets Jupiter and Saturn to each other in the year 7 BC. The other is the nova of 5 BC, which the Chinese saw, and recorded as occurring in the constellation Capricornus. Scholars still dispute which of these celestial events is the more likely to have found its way into the history of western civilisation.

Eventually, 1960 years on, American astronomer Robert Kraft discovered the reasons for a nova explosion, with the help of the Mount Palomar 5 metre (200 inch) telescope. His investigation centred on the remains of old novae. Statistical studies had already shown that novae explode more than once, their eruptions typically separated by 100,000 years. No one has seen an ordinary nova re-explode, but the related less energetic recurrent novae can explode as frequently as every couple of decades: Nova Pyxis, for example, erupted in 1890, 1902, 1920, and 1944. So Kraft's old novae were stars that will one day explode again.

He found that none of them is a single star. All consist of a white dwarf in orbit about a normal main sequence star, like the Sun but usually rather lighter in weight. The orbit is so tight that the white dwarf is actually drawing off gas from its companion. This gas shares the orbital motion of its parent, so it doesn't fall straight on to the foraging white dwarf. Instead, it forms into a spinning disc around the tiny star. Friction between the gas particles means that the gas gradually spirals inwards; and eventually it piles up around the white dwarf's equator. The gas is mainly hydrogen. When enough of it is compressed to the white dwarf's hot surface, the gas ignites as a cosmic-scale hydrogen bomb – a nova explosion.

The piled-up gas and the surrounding disc are blown out into space at tremendous speed, around 1000 kilometres (620 miles) per second. Despite the nova's ferocity, it doesn't seriously affect the white dwarf or its companion – the amount of matter ejected is only a thousandth that of the Sun. The system is swept clean temporarily, but soon the white dwarf starts pilfering gas again, and eventually another nova explosion will occur.

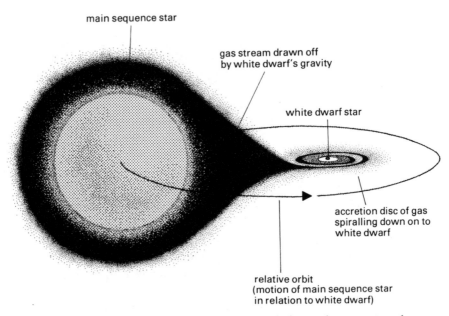

main sequence star

gas stream drawn off
by white dwarf's gravity

white dwarf star

accretion disc of gas
spiralling down on to
white dwarf

relative orbit
(motion of main sequence star
in relation to white dwarf)

15 A NOVA SYSTEM BEFORE OUTBURST *A nova system before outburst consists of a compact white dwarf star drawing gases off a normal, main sequence companion star. At this stage, the hot accretion disc surrounding the white dwarf is shining more brightly than either star. When enough gas has spiralled down on to the white dwarf's surface, the hydrogen gas in it explodes exactly like a hydrogen bomb. In this nova outburst, the system can flare up to a million times its previous brightness; but neither star is permanently affected and eventually the system settles down to this state again.*

Kraft's work not only explains novae, and their relatives the recurrent novae and the fainter dwarf novae, but also led to an understanding of X-ray stars when these were discovered. It was a fine example of why an astronomer needs expensive, large telescopes, in order to glean the maximum amount of light from faint objects. Britain's leading nova expert, Geoff Bath, calls Kraft's research 'not only the most important recent observational breakthrough in the subject, but also one of the major contributions to stellar astronomy by the 5 metre telescope at Mount Palomar.'

There is no reason why a neutron star too shouldn't be in close orbit with another star. The binary pulsar is orbiting another neutron star; but there must be many cases where the companion is an ordinary star. If they are close enough, then the neutron star can forage gases from the companion, just as the white dwarf does in a nova system. Here, too, the gas's orbital motion will spin it out into a disc surrounding the neutron star, and the gas gradually spirals down towards the compact star's surface, friction, all the while, raising its temperature.

The gas spiralling in towards a neutron star has further to fall though, because a neutron star is smaller than a white dwarf, and more important it falls through the region of intense gravitation near the star. As a result, it heats up much more than a nova's accretion disc. The central regions are at a temperature of over ten million degrees – and gas this hot emits copious amounts of X-rays.

Amongst the many types of X-ray source in the sky are several kinds where astronomers suspect the gas disc of a neutron star. The most obvious are the X-ray pulsars. Just like radio pulsars, these 'flash' radiation at very precise intervals, X-rays beamed out like a lighthouse beam by a rotating neutron star. Here the disc of gas 'robbed' from a companion is not settling down to the neutron star's equator. The pulsar's magnetic field channels the gas towards each magnetic pole, where it falls to the surface as a small *hot-spot* which emits X-rays powerfully into space.

The X-ray pulsars are undoubtedly old neutron stars, whose radio emission stopped long ago, and which have only sprung to life again because of the supply of gas from the companion. Being older, they should spin more slowly than radio pulsars. But the spiralling flow of gas on to an X-ray pulsar can change its spin rate. Some have been slowed right down so that they turn at a very leisurely rate – the pulsar X Persei takes fourteen minutes to turn once – while others have been spun-up until they turn as quickly as the young radio pulsars. The strong X-ray pulsar SMC X-1 is powerful enough to be detected even though it lies in another galaxy, the Small Magellanic Cloud, and it spins in less than a second.

The strongest X-ray sources in our Galaxy are probably similar star-systems. The prototype is called Scorpius X-1, and it was the first X-ray source (apart from the Sun) to be discovered in the sky. The team which found Scorpius X-1 had in fact launched their rocket in 1962 to look for X-rays from the Moon. They picked up X-rays, but careful analysis showed that they came not from the Moon but from a region in Scorpius 40° away in the sky. Better position measurements showed that the source can be seen optically as a faint bluish-coloured 'star' whose light flickers erratically – although we are actually seeing not the star but light from the hot gases surrounding the neutron star.

Sources of the Scorpius X-1 type emit ten thousand times as much energy in X-rays as the Sun produces as light and heat. Only the accretion disc of a neutron star could be so prolific – although this is an indirect inference because these sources don't give their nature away by pulsing.

Another intriguing class of X-ray sources is the X-ray burster type. These are quite weak X-ray emitters, which will suddenly intensify a hundredfold for a matter of just a few seconds before fading back to insignificance. The first burster was picked up by the Astronomical Netherlands Satellite late in 1975. Three months later, Walter Lewin and colleagues at the Massachusetts Institute of Technology discovered an amazing burst source, which they named the Rapid Burster. It fires off staccato X-ray bursts every minute or so, like a cosmic X-ray machine gun.

In the X-ray bursters, the neutron star's magnetic field keeps the incoming gas suspended above its surface. Eventually, however, enough gas builds up for its weight to crush the magnetic field, and the gas suddenly drops on to the star in a burst of X-rays.

Neutron stars may also produce the even higher energy radiation called *gamma rays*, which occasionally flash at us from the skies. These gamma ray bursts are even shorter and more intense than their X-ray counterparts, and their origin is still obscure – although neutron stars are the currently-favoured source.

The discovery of gamma ray bursts reads more like science fiction than fact. It shows that anyone dealing with high-technology equipment is liable to suffer from cosmic influences – and in the right circumstances, it may be not just a nuisance, but a clue to strange events happening in the Universe about us.

Since the Partial Test Ban Treaty of 1963, the United States Defense Department has kept a permanent vigil for unauthorised nuclear tests by means of satellites in Earth orbit. These satellites, codenamed Vela, monitor both the Earth and space for the sudden flashes of neutrons, X-rays and gamma rays that accompany a nuclear explosion.

Unfortunately, charged particles in space can trigger the detectors, so signals are only valid if more than one detector registers a burst. In 1973, Ray Klebesadel of the Los Alamos Scientific Laboratory was carefully checking through these 'spurious' events. And, like Jocelyn Bell's thorough but apparently uninteresting investigation of radio 'scruff' six years before, Klebesadel's diligence unexpectedly paid off. He found that two Vela satellites on opposite sides of the Earth were picking up sudden gamma ray bursts from space at the same time.

No one had predicted intense gamma ray bursts, and astronomers were puzzled by the news. There was no doubting the evidence, however, even though most astronomers had never even heard of the satellites that had made the discovery! With hindsight, it turned out that various astronomical satellites had also picked up these bursts. The experimenters had simply dismissed them as interference.

Now there is a network of gamma ray detectors spread through the solar system on a variety of spacecraft, including the German Helios 2 close in to the Sun and American and Russian Venus probes as well as Earth satellites. This international collaboration paid off on 5 March 1979, when the probes detected the most powerful yet burst of gamma rays. Combining the spacecraft results, astronomers have pinpointed this outburst to the remains of an exploded supernova (catalogued N49) in our nearest-neighbour galaxy, the Large Magellanic Cloud. Since this galaxy lies 180,000 light years away, the burst must have been extraordinarily powerful – thousands of times more energetic than a typical gamma ray burst.

Astronomers are still uncertain how gamma ray bursts are produced, but the March 1979 burst strengthens the idea that neutron stars are involved. After the burst itself, which lasted only a fraction of a second, the gamma rays continued at a much fainter level for a minute. And during this time they pulsed regularly every eight seconds. This pulsing behaviour, and the association with the gaseous remains of a supernova, are convincing evidence that a neutron star is responsible.

Neutron stars are the most compressed form of matter possible in the Universe. Although they were discovered by their radio pulses, neutron stars give away more secrets when studied by the very high-energy radiations, X-rays and gamma rays. As these new fields of astronomy expand – and gamma ray astronomy in particular is still in its infancy – astronomers will come to learn more about the bizarre nature of pulsars of all types, neutron stars in a wide range of astronomical situations.

As X-ray astronomer Harvey Tananbaum puts it, 'X-ray astronomy gives us a high leverage because it takes extreme conditions – high temperatures and strong [magnetic] fields – to produce them. X-rays point us to some of the most exciting events in the Universe.'

— 10 —

Black Holes

Young astronomers didn't escape the craze for badges in the 1970s. Their jangly decorations proclaimed 'quasars are far out', 'red giants aren't so hot', the inevitable 'astronomers do it at night' – and the always popular 'black holes are out of sight'. That's undoubtedly true of these strange invisible objects, but black holes were certainly not 'out of mind'. During that decade, astronomers came to realise that black holes are probably responsible for some of the most exciting events in the Universe. Now that black holes are an accepted part of the astronomical scene, scientists are busy exploring – theoretically at least – the strange effects of a black hole on its surroundings, and even calculating what lies inside the hole itself.

These calculations are very complicated, and some of the results seem to be pure science fiction – like the prediction that a black hole may be the gateway to another universe. But in essence a black hole is very easy to understand. It is simply the region around a very shrunken star where that star's gravity is so strong that it prevents light waves from escaping.

We are all familiar with the Earth's gravity, which anchors us firmly to our planet's surface. If we throw a stone upwards, gravity will slow it to a standstill, and then pull it back to Earth; but if we could enlist Superman's help to fling it upwards at a high enough speed, the stone would break the force of Earth's gravity altogether. The speed we need to give the stone is about 40,000 kilometres (25,000 miles) per hour – 11 kilometres (7 miles) per second. Astronomers call this speed the Earth's *escape velocity*. Even though it takes powerful rockets to escape from Earth in practice, our planet's gravity is pretty weak compared to some bodies in the Universe. Calculations show that a body's escape velocity depends both on the amount of matter it contains (its mass), and on how much it's compressed. So the massive Sun has an escape velocity of 620 kilometres (385 miles) per second; while a white dwarf star, which is just as massive but a hundred times smaller, has a much stronger gravitational pull and an escape velocity around 7000 kilometres (4300 miles) per second. As we saw in the last chapter, a neutron star is about the same mass again, but is compressed into a tiny sphere only 25 kilometres (15 miles) across: its unimaginably strong gravity would flatten an astronaut trying to land there, and crush him down on to the surface into a thin pancake only one atom thick. The escape velocity is around 100,000 kilometres (62,000 miles) per second – about one-third the speed of light itself.

We would only need to compress a neutron star a little more, and its escape velocity would become even higher than the speed of light. Such a star could be shining very brightly, but we wouldn't see it at all because the light waves aren't travelling fast enough to escape from its gravitation. Since no light escapes from the star, it appears completely black – blacker than anything else in the Universe.

According to Albert Einstein's theory of relativity, nothing can travel faster than light – the speed of light is the Universe's natural speed limit. Since light is not travelling fast enough to escape from the star, nothing else can get away from it either. Things can fall down on to the shrunken star; but they can never get away again. If a foolhardy spaceship commander took his craft too close to such a star, he would find himself in such a strong gravitational field that he could never escape from it. And he could never even tell other spaceships what had happened, because the gravitational pull would prevent light, radio signals and all other radiations from getting away. His wiser colleagues who've kept at a safe distance would see him disappear into the small black 'star', and vanish forever. The star's gravitational pull creates a 'hole' in space, which objects can fall into, but never come out again. It is a 'black hole'.

Black holes sound like science fiction – yet they are almost certainly science fact. Our Milky Way Galaxy alone must harbour about a hundred million black holes, drifting about unseen in space. They are the final, collapsed remains of heavyweight stars – ironically enough, the blackest things in creation are the corpses of the most brilliantly shining stars. As we saw in Chapter 8, a star's own gravity is continuously trying to shrink it in size, and it's only the force of radiation welling up from the centre that prevents a star like the Sun from collapsing. When the Sun's 'fuel' is eventually spent, it will shrink to become a white dwarf star, where the electrons are jammed together so tightly that they stop the collapse. But there's a limit to what even these subatomic particles can stand. The cores of heavier stars eventually end up as neutron stars, where the smaller-sized neutrons are packed together. If you try to assemble a white dwarf more than 44 per cent heavier than the Sun, its electrons can't stand the pressure: they will merge with the protons to form neutrons and the white dwarf will collapse to become a neutron star. And there's a natural weight limit for neutron stars, too, at three times the mass of the Sun.

But we know there are ordinary stars much heavier than this in the Galaxy (Chapter 8). A star fifty times heavier than the Sun may blow off most of its matter in a supernova explosion, but it still has a core weighing as much as ten Suns. Its own gravity shrinks the core, smaller and smaller. It's too heavy to stop collapsing and form a white dwarf or a neutron star, and it must shrink to the point where its escape velocity becomes equal to the speed of light. It has become a black hole.

In fact, the calculations show that the star's core won't stop shrinking even then. Once its gravity has formed the black hole, it will continue getting rapidly smaller and smaller within the privacy of its black hole, until it's virtually no size at all. The matter of ten Suns will be compressed into a point far, far smaller than an atom – in theory, into literally zero size, a mathematical point. Theorists call this tiny, infinitely compressed matter at the centre of a black hole a *singularity*. It's surrounded by the black hole itself, a sphere of empty space a few kilometres across, where gravity is so strong that nothing can escape.

Theoretical astronomers are convinced that massive stars must end their lives as black holes, and as we've seen, they can even estimate roughly how many there should be in our Galaxy. But *finding* a black hole is another matter. And unless we can discover a black hole, we can't state that black holes do actually exist. As one astronomer has put it, the search for black holes is like 'looking for a black cat in a coal cellar on a dark night – except worse'.

We can't hope to 'see' a black hole itself: it can't give out light, radio waves or any other radiation. If we could get close enough, we'd see it as a dark silhouette in front of the distant stars. But we'd need to approach within a few hundred kilometres, and that would mean knowing where to find it in the first place, because even the nearest black hole is likely to be at least fifty light years away – 500 million million kilometres (310 million million miles).

But it's quite easy to locate your cat in the cellar if it's having an altercation with another cat, especially if the other cat is easily visible. We saw in the last chapter how a neutron star can rip gases away from a companion star, and how this gas swirls around the neutron star in a hot, swirling disc as it spirals inwards under the pull of the neutron star's gravity. This hot disc of gas generates copious amounts of X-rays. Satellites like the Einstein Observatory, orbiting above the Earth's atmosphere, can pick up these radiations from space, and pin-point the location of these distant X-ray sources. If a black hole has a companion star, it too can cannibalise its neighbour; and the disc of hot gases around the hole will send out prodigious amounts of X-rays before the gases disappear down the throat of the black hole.

So the X-ray sources which appear in the astronomers' catalogues mark the hiding places of either neutron stars *or* black holes. Some of them produce regular pulses of X-rays, the hall-mark of a neutron star 'pulsar'. Astronomers need very strong evidence before they can conclude that any of the others are the work of black holes – whose existence is still to be proved – rather than neutron stars, which are a firmly accepted beast in the astronomical menagerie.

The X-ray source known as Cygnus X-1 seems to provide that evidence.

This is a double star system, and optical astronomers can see one of the pair, a very bright star called HDE 226868. Its partner is invisible, and must be small because it is surrounded by a hot gas disc which produces the X-rays (see page 113). But is it a neutron star or a black hole? The vital clue comes from 'weighing' this compact unseen star – a task not as difficult as it seems, because astronomers can deduce its mass from its effects on the bright companion. The small star turns out to weigh as much as ten Suns. This is far above the natural weight limit of three Suns for neutron stars; and astronomers have concluded that the compact 'star' in the Cygnus X-1 system is a heavy black hole.

It's an indirect argument, though, and many astronomers have slight reservations in case there's some weak link in the argument that hasn't yet been spotted. Rather than saying 'we have proved that a black hole exists', most astronomers would go along with the cautious approval of black hole theorist Stephen Hawking when he says he's '90 per cent sure that there is a black hole in Cygnus X-1'.

Unfortunately, there aren't any other X-ray sources where the invisible star has been weighed and come out over the limit for neutron stars. Astronomers do however have a short-list of other black hole suspects, which have raised suspicions because of peculiarities in their X-ray output. Some of these 'stars' produce light in a peculiar way, too, an oddity which becomes particularly obvious when the light is spread out into a spectrum of wavelengths. One such suspect is an object called Circinus X-1, and it's interesting not so much for itself but because it led astronomers to discover what's probably the most peculiar 'star' in our Galaxy.

Circinus X-1 lies amidst the extended gas cloud thrown out from a supernova explosion, and astronomers naturally surmised that this supernova's core collapsed to become the black hole in Circinus X-1. So they started to search in the central regions of other old supernova gas clouds to look for evidence of similar lurking black holes. Radio astronomers at Cambridge found that radio waves were coming from the centre of one old supernova remnant in the constellation Aquila. This in itself didn't reveal very much about what was going on. What astronomers needed was a spectrum of the light coming from whatever was there.

At this time, in 1978, British astronomers David Clark and Paul Murdin happened to be using the Anglo-Australian Telescope (AAT), the telescope described at the beginning of Chapter 1. They had half-an-hour to spare from their intended observations, so they turned the telescope to the position of this radio source in Aquila, to try and find the object, and take its spectrum. As Clark and Murdin reported in *New Scientist* 'suspecting any candidate to be, like Circinus X-1, extremely faint, and not keen to risk the telescope's sensitive apparatus by shining too much light at it, we initially skirted round the brightest star we found at the radio position. We took spectra of a couple of fainter stars in the field but they were boringly ordinary. Obviously the brighter star was worth a careful look. As we nudged the telescope on to the image of the star, and before we actually began to take data, an optical spectrum remarkably similar to that of Circinus X-1 revealed itself on the computer displays, to the delight of a jubilant pair of astronomers in the AAT control room.'

This star, SS433, turned out to be stranger than anyone could have expected. American astronomer Bruce Margon looked at its spectrum week by week and discovered that some of the 'lines' were changing their wavelengths continuously, marching along the spectrum. Eventually they stopped and moved back the other way. Although these changes were totally perplexing at the time, astronomers now think that this light comes from fast jets of gas that are swivelling around, like a lawn sprinkler. The wavelengths change by the Doppler effect (see page 130), shortening when a jet is pointing towards us, and lengthening when it's directed away.

There's still some dispute over what SS433 actually *is*, but many astronomers believe it consists of a heavy giant star and a black hole in orbit around one another. As in the Cygnus X-1 system, gas from the giant star is streaming off into a disc around the black hole; but in SS433, some gas is boiling off either side of the disc, as two oppositely-directed jets. The gas making up these jets is travelling faster than anything else known in the Galaxy – at a quarter the speed of light! The disc is slightly tipped, and it wobbles like a toy top, swivelling the jets around as it does so.

Exciting as such discoveries are to astronomers, black holes have had rather a bad press outside the scientific community in recent years. Sensationalised accounts speak of the dangers of black holes 'sucking in' unwary spacecraft – or even sneaking up and swallowing the Earth. In reality, however, black holes are not likely to be a cosmic danger. They can have very little influence on life on Earth, nor on future interstellar journeys.

One widespread – and mistaken – idea is that a black hole is like a cosmic octopus: feel just the slightest touch of its outspread gravitational tentacles, and you are doomed to be sucked into the hole. This is quite simply not true. The point of no return is the edge of the black hole itself. If you venture inside, certainly you can never return; but in the region outside the hole its gravitational pull is no more remorseless than any other body's. For example, if the Sun suddenly collapsed to become a black hole, the Earth wouldn't be sucked in, any more than it is 'sucked in' by the Sun's actual gravity. At present, the Earth's speed in its orbit ('centrifugal force') prevents it from falling in to the Sun. If the Sun collapsed overnight, the resulting black hole would exert exactly the same gravitational pull on the Earth, and our planet would simply continue to follow the same orbit. The only difference we'd notice is that the 'day time' would be as black as night.

So a black hole is a danger only if you get really close to it. And the chances of a black hole running into the Earth really are negligible. Because black holes are much smaller than stars, and much rarer in space, it's

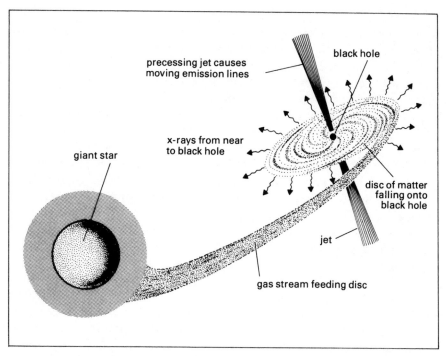

16 SS433 – A BLACK HOLE AND A HEAVY GIANT STAR? *The fast jets of gas in the strange SS433 star system may come from the sides of a disc of gas spiralling into a black hole. In this theory, the black hole is drawing gas from a giant companion star and the jets consist of gas streams 'boiled off' by the very high temperatures reached in the innermost parts of the disc, just outside the black hole itself. The black hole and the giant star orbit around one another every 13 days; the gravitational pull of the giant on the inclined disc makes the disc precess ('wobble' around) with a period of 164 days, and as a result the jets too precess around with this period, like an eccentric garden lawn sprinkler.*

actually a million million times more likely we'll be hit by a star than by a black hole – and even that is an almost impossibly-rare circumstance. Spaceships will be in no real danger either. After all, the Earth itself is a spaceship, travelling with the Sun through our Galaxy, and a small man-made starship is less likely to encounter a ten-kilometre (six-mile) black hole than is a whole planet.

And that brings us to another popular idea about black holes, the possibility of actually using black holes for space travel. According to science fiction stories – and even a Walt Disney film – we can travel into a black hole in our Universe, and emerge unscathed into another, totally different universe. Perhaps black holes might even connect with a white hole somewhere else in our Universe, so we could use this 'wormhole' to cut corners in space-travel, to traverse vast reaches of space in virtually no time at all.

These ideas read like no more than science fantasy, but they do have a solid foundation in theory. In an ordinary black hole, any traveller foolish enough to venture inside would find himself pulled right down to the central tiny singularity – and there he would be crushed to zero size. But the theory of black holes tells us that they can rotate. It's a little difficult to imagine a 'hole' spinning around, although that is the technical description. In practice, it means that the central singularity will not pull a space-traveller directly towards the centre. Its gravity will continuously pull him sideways too, so that he spirals inwards. And once he's well within the hole, down near the singularity, his sideways speed can give him enough centrifugal force to balance the gravitational inpull of the singularity.

He now has a respite, safe from being crushed into the singularity, although he can't get back to our Universe again. But theory shows that the safe haven may be connected to a similar region in another universe; a region which is however the centre of a *white hole*. This is the exact opposite of a black hole. Just as matter is dragged inwards once it's inside a black hole, so matter inside a white hole is inexorably pushed *out* into surrounding space. Once our astronaut traverses the short 'tunnel' from one interior to the other, he will be jettisoned willy-nilly into the other universe.

Although these calculations have inspired science fiction writers, a deeper look at the interior workings of black holes has shown that these tunnels just won't work. The trouble arises at the outer edge of the 'safe haven', where centrifugal force balances gravitational inpull. This is called the *inner event horizon*, to distinguish it from the *outer event horizon* which marks the outer boundary of the black hole. Detailed calculations show that some apparently minor effects (such as a build-up in radiation which has fallen into the hole) build up at the inner event horizon, to create another singularity – in effect to build an inpenetrable barrier. A real spaceship would hit this barrier at the inner event horizon and be crushed to nothing, before it could even reach the safe haven within, let alone the 'tunnel'.

Our present knowledge tells us that we cannot use black holes to help us traverse the cosmos. But there are still a lot of problems to be solved in the theory of black holes. And one of the most important is the question of 'cosmic censorship'. We've seen that a rotating black hole has two parts: an inner safe region, containing the singularity at its centre and bounded by the inner event horizon; and an outer region between the inner and outer event horizons. Anything in this outer region is pulled inexorably inwards. This region acts as the one-way valve that allows matter and radiation into the hole, but prevents it from getting out. The faster the rotation of the singularity, the greater the centrifugal force it gives to anything in the region around. The inner event horizon – where centrifugal force balances gravity – moves outwards. And so the outer, one-way region of the hole shrinks.

In the extreme, the hole could spin so fast that the inner event horizon

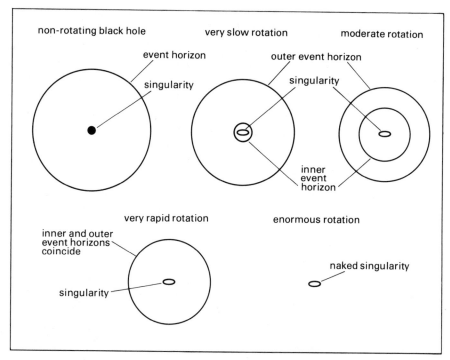

17 ROTATING BLACK HOLES AND NAKED SINGULARITIES *A rotating black hole is surrounded by two event horizons: between them, any object or beam of radiation is forced to follow an inward path, but once it is within the inner horizon it can move freely. The size of the inner horizon depends on the hole's speed of rotation, and if the hole is rotating fast enough, the inner horizon can become as large as the outer horizon. The central singularity is then no longer concealed from the surrounding Universe by an outer event horizon: it is a* naked singularity.

has moved outwards to coincide with the outer event horizon. There is now no one-way region. Spaceships can travel as they please, right in to the singularity, and then away again, back home. And the singularity is exposed: it is a 'naked singularity'. Radiation from the naked singularity escapes to space, so we can see it; and anything material that the singularity emits can get away too – because, crudely speaking, the singularity can supply a centrifugal force greater than its gravitational inpull.

This may all seem highly theoretical, but many scientists are worried about the possibility of naked singularities just lying about in space. The problem is that singularities wouldn't obey the laws of nature. They are pure-bred anarchists. A singularity could create matter from nothing, for example, and suddenly spew out a galaxy, an elephant – or something we just cannot imagine. And it could affect a spaceship innocently passing by – perhaps swopping over the heads on the crew members, and turning the ship into a hollow pumpkin.

Leading black hole theorist Stephen Hawking calls the possibility of naked singularities 'a great crisis for physics, because it means that one cannot predict the future'. British singularity expert Paul Davies says 'when naked singularities are involved, one is free to speculate that anything is possible'. One single naked singularity far away in the Universe could wreak havoc on the Earth by causing weird and unpredictable events.

Many scientists, however, believe that Nature does not allow naked singularities to exist. Black hole pioneer Roger Penrose has asked the question 'does there exist a "cosmic censor" who forbids the appearance of naked singularities, clothing each one in an . . . event horizon?' Theorists are busy calculating whether our theories do indicate that an event horizon always exists around any kind of singularity. Most theorists believe that some kind of cosmic censor must exist; if it doesn't, then our world – and the rest of the Universe – is wide open to the worst of all possible cosmic threats, the maverick, supernatural power of a naked singularity.

Astronomers in general are not too perturbed, however, for there is no evidence in the skies for such a beast as a naked singularity. They have enough of a task to track down the black holes which clothe a normal decent singularity. As well as black holes a few times heavier than the Sun (in star systems like Cygnus X-1), there could be in principle black holes of any other size, their size depending on the amount of matter contained within them.

If we shrank the Earth down to zero size, for example, it would be surrounded by a black hole a couple of centimetres across – the size of a large marble. More realistically, a huge mass of gas at the centre of a galaxy might coalesce to make a black hole as heavy as several thousand million Suns. As we'll see in Chapter 15, astronomers do have evidence for such huge black holes in some galaxies. Their gravity may be the force behind the awesome power of the distant quasars.

At the other end of the scale, the Universe may contain countless numbers of very tiny black holes, created during the Big Bang in which our Universe was born. Each of these would weigh perhaps a thousand million tonnes – the mass of a small asteroid – with a size smaller than an atom. And oddly enough, such black holes should be exploding – destroying themselves in a burst of radiation which would rival a million megatonne bomb!

This prediction seems to contradict all that we know about black holes – that they draw matter and radiation inexorably inwards. But it's undoubtedly true. And even more remarkable than the prediction itself, the original complex calculations were all done by 'mental arithmetic' – in the head of a brilliant young scientist who is severely crippled by a muscular wasting disease.

Stephen Hawking was already a respected physicist at Cambridge in his mid-twenties, when his disease confined him to a wheelchair. But he hasn't lost his sense of humour – nor his mathematical genius. As he turned thirty, he began to investigate what happens in the space around very small black holes. Without the use of his hands, Hawking had to rely on friends to turn the pages of the books and scientific papers that he read – and couldn't put his own long calculations down on paper. He had to work it all out in his head.

Hawking reached the remarkable conclusion that black holes radiate away energy. This doesn't contradict the fact that nothing can escape from a black hole, because Hawking's radiation comes from the region of strong gravity *just outside* the black hole itself. It's not possible to describe his result accurately in everyday language, but one way of putting it goes as follows. The strong gravitational field just outside the black hole can split up empty space into positive energy and negative energy. The positive energy turns into radiation and subatomic particles like electrons, and these shoot off into space. The negative energy flows *into* the hole, right down to the singularity at the centre. Here it cancels out some of the matter which originally went to form the hole. The upshot is that the singularity gradually becomes lighter in weight, and as it does so, the hole shrinks in size. All the time, the loss in weight is exactly balanced by the radiation and sub-atomic particles simultaneously appearing outside the hole.

This process is slow to begin with, but speeds up as the hole becomes smaller, until it finally explodes. If the Big Bang produced black holes weighing only a thousand million tonnes, they should now be reaching the end of their lives and should be exploding around us at the present time.

Astronomers haven't yet found any sign of exploding black holes, but that doesn't mean Hawking is wrong. It simply tells us that the Big Bang didn't produce many of these 'mini black holes' after all. Indeed, many other theorists have checked Hawking's calculations thoroughly, and agree with his conclusions. A leading British expert on gravitation, Dennis Sciama, describes Hawking's published account as 'one of the most unexpected and beautiful papers in the whole history of physics'. And there's an even more flattering comparison. Our present theory of gravitation was formulated by the great Albert Einstein during the First World War, and several scientists have remarked that Hawking's amazing brain-work has improved our knowledge of gravity more than anything since Einstein's own work.

Most scientists now accept the idea of black holes quite seriously. But it's unlikely that future astronauts will ever explore the interior of a black hole – and even if they did, they could never send out signals to tell us what they had found. They would simply disappear from our Universe – possibly to reappear in another universe, more likely to end in oblivion. The only way we can explore black holes in reality is by patient and tedious calculation, with computer, pencil and paper, or – as Stephen Hawking has shown – simply with the power of the human mind.

— 11 —
The Milky Way

To the ancients, the Milky Way – that hazy band of light which spans the sky on a clear summer's night – held no mysteries. If anything, their explanation for it was somewhat down-to-earth. They told the legend of Hercules, one of Jupiter's countless illegitimate babies, who was desirous of immortal life. Unfortunately for him, his mother was but a mere mortal; a fact which should have put paid to such aspirations for good. But Jupiter took pity on the child. Suckling at a goddess's breast, he said, would guarantee immortality – and then directed the infant towards the bedchamber of his wife, the goddess Juno. As Juno slept, Hercules crawled and toddled across the floor, his eyes fixed on her gently rising and falling breasts. But just as his goal was almost within reach, Juno – no doubt tickled – suddenly woke up. Instead of going into Hercules's mouth, a stream of goddess's milk spurted across the sky as Juno arched backwards in her shock – and thus was created the Milky Way.

Despite the ancient's delightful certainty, only this century was the Milky Way's nature fully realised. Ever since Galileo first turned a telescope to the sky in 1609, it was known to be made of stars, but astronomers were unsure how these distant and apparently crowded stars could be reconciled with those they could see scattered over the whole sky. Eighteenth-century philosophers like Immanuel Kant and Thomas Wright hazarded commendably accurate guesses, but the astronomers of that time busied themselves elsewhere with the pressing problem of confirming the universal nature of Newton's law of gravitation. William Herschel was the one astronomer of those times to tackle the question scientifically, by painstakingly counting the stars visible through his great telescope in order to see just how they were distributed through space. He even hit on the correct explanation of the Milky Way, but it was not accepted by the nineteenth-century astronomers.

By the start of the twentieth century, many strands of evidence had knitted together to show that the Milky Way band is just an edge-on view of the distant stars making up our home Galaxy. The discovery of millions of other galaxies spread through space – each a separate star-island with a membership running into millions – also helped astronomers to gain a perspective on our own.

With hindsight, the broad-brush picture of our Milky Way Galaxy is surprisingly simple. It is a disc-shaped system of stars, gas and dust wheeling slowly in space. From the side it would look like a spindle, widening towards the central bulge – or, in the words of one astronomer, 'just like two fried eggs clapped together back to back'. A bird's-eye view would reveal the beautiful Catherine-wheel shape of a typical spiral galaxy, so wide that a ray of light would take 100,000 years to cross it. Our intergalactic bird would be very hard-pressed to spot the Sun, for it is just one out of the Galaxy's 100,000 million stars, situated in an unceremonious position in the galactic suburbs some two-thirds of the way from the centre. We live in the outer part of our Galaxy's disc, and it is this positioning which gives rise to the band of the Milky Way in our skies. Looking into the thickness

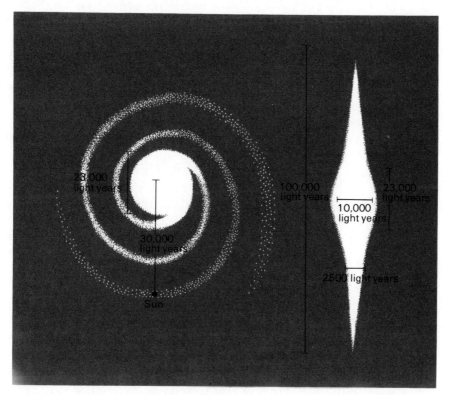

18 THE MILKY WAY *Our Milky Way Galaxy is a vast spiral island of some 100,000 million stars slowly turning in space. Its plan view (left) and edge-on appearance (right) are shown here. The nature of the Milky Way was not fully realised until this century. Ever since the time of Galileo it was known to be made of stars but astronomers were unsure how these distant and apparently crowded stars could be reconciled with those they could see scattered over the whole sky. Even now slight uncertainties about our Milky Way's extent still remain – particularly with regard to its surrounding spheroidal halo, and the number of spiral arms it contains.*

127

of the disc all around us, there are stars as far as we can see, tailing off into a misty blue. These distant stars *appear* crowded, but it is only an effect of perspective. When we look above or below the disc, the stars quickly run out: beyond is just empty space.

In many ways, it's easy to understand why the discovery of our own Galaxy came so late. Our very location inside the Galaxy means that it is hard to see the wood for the trees. And picking out its structure requires the measurement of distances to the stars, first successfully accomplished by Friedrich Bessel as recently as 1838. Even a century and a half later, measuring distances – and to some extent, probing our Galaxy's geography – are both enterprises which stretch astronomers to the limit.

Only the closest stars are accessible with the *parallax* method that Bessel (and his fellow-astronomers) used. It's a method familiar to anyone who has watched a surveyor at work with a theodolite, and it relies on the fact that a relatively nearby object, when viewed from two different vantage points, appears to shift slightly against a fixed, distant background. The parallax 'jump' is largest for nearby objects and becomes progressively smaller for objects further away – try this out by holding your finger close to your face and looking at it with each eye alternately; now repeat the manoeuvre with the finger held at arm's length. Finding the distance to the object, whether a finger or a distant hill, is a simple question of geometry – you need to know the spacing of the two vantage points, and measure the angle of the shift due to parallax.

The nearest stars are so remote that even measurements made on opposite sides of the Earth would not reveal a parallax. Instead, astronomers need to use the widest baseline available to them: the 300-million kilometre (186-million mile) span of the Earth's orbit about the Sun. The theory is simple. All that is required are photographs of the target star taken six months apart (with the Earth on opposite sides of its orbit) so that its parallax shift will show up against the fixed pattern of the background stars.

Of course, the practice is nothing like as easy. The annual parallax of the nearest star, Proxima Centauri – a mere 4.3 light years away – is three-quarters of a second of an arc. In detecting this shift, astronomers have done the equivalent of measuring the diameter of a British 1p coin at a distance of over 2 kilometres (1.2 miles)! And this is the biggest shift available: astronomers routinely measure parallaxes a tenth or even a hundredth of this. Even so, inaccuracies jostle to leap in and sabotage these measurements. Are the background stars really fixed – or do any show a small parallax? Has the target star shifted slightly because of its own real motion (*proper motion*) through space? If measurements were made in different seasons, have temperature changes made the telescope's optics sag? What about varying atmospheric conditions? A good parallax-program takes years, and needs at least fifty photographs of each star, analysed

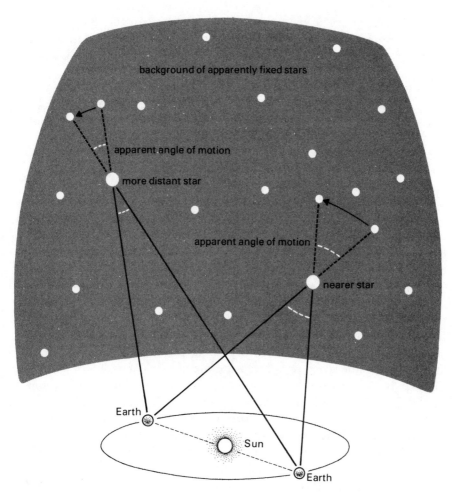

background of apparently fixed stars

apparent angle of motion

more distant star

apparent angle of motion

nearer star

Earth

Sun

Earth

19 THE PARALLAX METHOD OF FINDING A STAR'S DISTANCE *Astronomers use a surveyor's technique to measure distances to the nearest stars. Using the Earth's 300-million-kilometre (186-million-mile) wide orbit as a baseline, they measure the tiny angle of the 'parallax shift' of a nearby star against the distant background, by viewing the star from opposite sides of the Earth's orbit. Knowing this angle, and the diameter of our orbit about the Sun, it is then simple trigonometry to calculate the star's distance. However, the method is not nearly as simple in practice. Parallax shifts for more distant stars (left) are very small and extremely hard to detect; and even in the case of close stars, it needs many years of patient measurement and remeasurement to eliminate possible errors.*

under the same conditions, to obtain the necessary accuracy. This isolated, painstaking work, using old refracting (lens) telescopes – the best suited to programs like this – is very much 'astronomy of the old school'. But because distances are so important in astronomy, it is still of enormous value. Even so, there is a limit. Beyond 300 light years, the parallax method breaks down, for the errors of measurement become comparable to the measurements themselves. Distance measurements to a few thousand stars within 300 light years do not penetrate far in a Galaxy some 100,000 light years across.

Fortunately other, less direct, ways of measuring distances come to our rescue. Nearby star clusters (see Chapter 8), whose members all move in the same direction through space, offer one means. By analysing the spectrum of each star in the cluster, an astronomer can find its *radial velocity* – its speed directly towards or away from us. He does this by measuring the tiny *Doppler Shift* in the spectral lines caused by the star's actual motion: a shift towards the blue (shorter wavelengths) if the star is approaching us, with its lightwaves compressed in our direction; or a shift towards the red (longer wavelengths) if the star is moving away and its waves trail behind it. Then, by comparing these velocities with the actual motion of the whole cluster across the sky, and putting in a correction for the real direction in which the cluster moves through space, an astronomer can work out how far away the star-group lies.

Involved though it may sound, the 'Moving Cluster Method' really does yield very accurate distances. Its precision lies in the fact that each cluster contains hundreds of stars, and every one is individually measured, and its power rests in its ladder-like ability to push out to greater and greater distances. Imagine a very remote cluster, out of reach of the Moving Cluster Method. Its stars are the same types as those in nearby clusters, but dimmed by distance. Now compare the brightness of these stars with the real brightness of the members of a cluster whose distance is known. Result: a very accurate distance to the 'mystery' cluster, achieved by comparing hundreds of stars of different types and using the 'inverse square law' of radiation which determines how brightness is related to distance. In fact, the distance to any star can be roughly estimated in a very similar way, by comparing its apparent brightness with the actual luminosity a star of that temperature and type (giant or dwarf) ought to have according to the Hertzsprung-Russell diagram. Misleadingly, this method is sometimes called 'Spectroscopic Parallax', although it is completely unconnected with the familiar 'trigonometrical' parallax.

As well as the desire to probe the Galaxy, astronomers measure distances with an eye to finding and calibrating 'standard candles' – stars whose special brightness or obvious variability make them easy to pick out even beyond our Galaxy. Cepheid variable stars are among the most tried and trusted beacons. These celestial searchlights are thousands of times brighter than the Sun, but it is their regular variability which makes them such excellent distance probes. By calibrating Cepheid variable stars whose distances can be measured more-or-less directly, astronomers have found that their periods of variability (which range from about two to forty days) are very strongly related to their intrinsic brightness. A relatively faint

Cepheid will go through its brightness changes in only a few days, while the most luminous take several weeks to go through their cycle of pulsations (which depends on the size of the star). So obvious are these stars that an astronomer can measure the period and apparent brightness of a target star; read off its intrinsic luminosity from his calibration graph and then work out its distance.

Lest any errors creep in, there is a veritable armoury of back-up methods available to the astronomer involving other types of standard stars. Among these are RR Lyrae variable stars: stars 100 times brighter than the Sun which take a day to complete their cycle of brightness. Red supergiant stars all have roughly the same brightness, many thousands of times greater than that of our Sun; and most luminous of all are the brilliant blue supergiants, twenty-thousand times brighter.

Of course, errors can, and do, find their way in. At a sweep, the distance scale of the Universe changed by a factor of two in 1952 when Walter Baade, following a critical series of observations with the 200-inch Palomar telescope (now called the Hale 5-metre) during the wartime blackout, announced that there were two parallel classes of Cepheid variables. Each class had a separate period-luminosity relationship, but unfortunately, astronomers had generally been measuring the wrong stars. Also, distance-measurement would be far easier if space were empty for it's not just distance which dims the light of remote stars; between the stars, in 'empty' space, lurks an unevenly-distributed pall of cosmic soot. We will return to these ubiquitous dust-grains later in this chapter.

In spite of all these difficulties, astronomers have been able to build up a pleasingly consistent picture of our Galaxy's structure. Even so, it appears that there may well be more to it than meets the eye – quite literally. Take, for instance, the *galactic halo* – a vast, spherical and apparently empty region 300,000 light years across which completely surrounds the disc of our Galaxy. It's the ghost of the former gas-cloud which was to collapse and give birth to our present Milky Way, and until recently, it was thought to be as insubstantial as the next ghost. But now there are some good, if controversial, reasons for thinking that the disc of our Galaxy wouldn't rotate in the smooth, regular way we observe *unless* it is stabilised by a halo whose mass is ten times that of the disc. To put it another way: there is a very respectable camp of astronomers who believe that we only see one-tenth of our Galaxy's total mass and that the remaining nine-tenths resides, in some hidden form, in the halo.

On the surface, the halo really does look sparse. Its obvious distinction is that it lays claim to the Galaxy's first citizens; the stars which formed perhaps 13,000 million years ago when the young gas cloud was just starting to collapse. Now these stars drift old, red and lonely through the seemingly yawning gulfs – their one compensation being the magnificent view of the Milky Way's face-on disc in their skies. Many of the halo stars are clumped into dense globular clusters, tight balls of sometimes half a million stars measuring only a few dozen light years across. There are over a hundred of them, some (like Omega Centauri and 47 Tucanae in the southern hemisphere, and the northern hemisphere's M13 in Hercules) visible without optical aid. They float forlornly around the halo like goldfish trapped in a huge bowl, but just occasionally one will get sufficiently far from the Galaxy's gravitational maw that it can break away into the unknown territories of intergalactic space.

What then could possibly make up the missing mass? It would be convenient if it took the form of gas left over from our Galaxy's formation; but things are seldom that simple, and to all intents and purposes, the halo is gas-free. For a possible explanation, we need to turn to objects which give out very little light or indeed *any* radiation. At the moment, astronomers have their money on three possibilities. The halo may be chock-a-block with neutrinos whose tiny (but, as yet, unproven) mass will have a cumulative effect. It could (although it is difficult to see why) be filled with black holes. And it's just on the cards that it is packed with underweight stars – 'mini-stars or maxi-Jupiters' suggests one scientist – whose mass at birth was insufficient to trigger the fusion reactions which would make them shine. But all these ideas are controversial. Many astronomers are quite happy to leave the halo as the empty graveyard it appears to be and concentrate instead on that part of our Galaxy where the action is based today – in the disc.

Surrounding the old central bulge, the disc is our Galaxy's newest neighbourhood. Its star population contains a fair sprinkling of older inhabitants, but many of these are relegated to its extreme outer fringes, well above or below the galactic plane. In fact, the whole distribution resembles a series of tidemarks, with stars marking the collapse of our Milky Way from gas cloud to thin disc. And it's here, in the very innermost plane of the disc – where the Sun and planets are privileged to lie – that our Galaxy is still forming. Here, the spaces between the stars are filled with invisible, tenuous gas, the raw material of future stars. There are only half a dozen atoms in a matchboxful on average, but in time these isolated atoms – mostly virgin hydrogen, but increasingly atoms of carbon, nitrogen and oxygen processed in stellar furnaces and ejected into space – will clump together in huge clouds to produce the next generations of stars. And as we saw in Chapter 8, the products of starbirth are all around us. Our skies are bright with young nebulae, sparkling star clusters and dazzling blue supergiant stars, whose profligate use of fuel sentences them to short, but happy lives.

But there's more to these youngsters than good looks. Measure their distances and still more of our Galaxy's geography is revealed. These young objects are not spread throughout space, but bunch together in three distinct bands – one at about the Sun's distance from the galactic centre, the others to either side – and these, without doubt, mark the positions of the three nearest spiral arms in our Galaxy. The Sun is a member of the Orion arm, and it is flanked by the (inner) Sagittarius arm and the (outer) Perseus arm. But without a knowledge of their distances, all these stars and gas clouds seem to merge together in our skies to make up the band of the Milky Way.

Simply by using our eyes – extended to their limits by telescopes and cameras – we have found that our Galaxy is a spiral. Perhaps it looks like one of those beautiful 'catherine wheel' photographs of Galaxies beyond our own; or are we speaking too soon? Because of the tremendous absorption of light caused by the interstellar dust-grains, our present picture is limited to a circle, centred on the Sun, only 10,000 light years across. Can we be sure that our own neighbourhood is typical? And are there any ways of combating the invidious effects of the cosmic soot?

At first sight, the regions of 'soot' looked like a veritable blessing. Dark clouds of dust-grains (like the Horsehead Nebula in Orion or the southern hemisphere's 'Coalsack') looked so much like empty starless voids that many nineteenth-century astronomers thought them to be 'tunnels' through which we could peer vast distances into space. Ironically, these dark clouds are among some of the densest concentrations of dust in the Galaxy: regions inside which the light from currently-forming stars is dimmed over a thousand million times. Obscuration generally isn't quite as bad as that: the problem is that it is unpredictably uneven. When estimating distances, astronomers adopt a rough figure for dimming by dust of 1 magnitude (a factor of $2\frac{1}{2}$) per 3000 light years. But this is an average, and it could be wildly out in certain directions – particularly along the central plane of the disc where the structure of the Galaxy is, predictably, at its most interesting.

Although they may not be certain about the exact extent of the obscuration by the grains, the majority of astronomers agree about their nature. It seems that each grain is a tiny speck of silicate rock, less than a micron (that's $\frac{1}{10,000}$ of a centimetre) across, coated by a thin layer of ice. Together, the grains make up only 0.1 per cent of the Galaxy's mass, but their effect on our perception of it is devastating. However, lest we risk being too unkind to the grains, we must remember the invaluable role they play in cocooning young stars from disruptive radiations as they form, and how – ultimately – they knit themselves together into planets like our own.

The most likely initiators of this Galaxy-wide celestial smokescreen are ultra-cool red giant stars, with temperatures of less than 3000 °C (5400 °F). Grains are believed to form like soot on their surfaces, and are eventually driven out into space by powerful stellar winds.

A very small number of astronomers disagree with this picture. Sir Fred Hoyle and Chandra Wickramasinghe have come to the startling conclusion that the spectra of the grains resemble instead certain complex organic ('life-forming') substances such as cellulose. They go on to suggest that the role of the grains may be of more direct importance to us than we could ever imagine: in short, they are the 'seeds' which initiate and spread life throughout the Galaxy. Understandably, these ideas are highly controversial, and we will look at them more closely in Chapter 13.

As far as the fog-bound astronomers on Earth are concerned, the grains have one very big thing in their favour – they are small. So, while they are impenetrable obstacles to light waves, whose wavelengths are roughly the same size as the dust-grains, they are practically invisible to longer wavelength radiation. An astronomer 'tuning-in' to the Galaxy at long infrared and radio wavelengths will not 'see' any dust-grains at all. And so, with modern techniques, we can 'look' straight across the Galaxy. No longer do we find ourselves in the situation which leading galactic structure expert Bart Bok describes as 'trying to map a large city from a suburb on a misty day'.

But even so, our position in the Galaxy gives us a worm's-eye view of that 'city'. Radio astronomers still have to face the complicated task of disentangling dozens of signals piled atop each other along the same line of sight, and that's one reason why the exact shape of our Galaxy is open to debate even now. At least Nature has co-operated by providing a strong, clear signal to help the job of mapping. It comes from the cold hydrogen gas which pervades the space between the stars. Although invisible to us, the signal it gives out at a precise wavelength of 21.1 centimetres (8.4 inches) is a veritable beacon to radio astronomers. Its real selling-point as a mapping signal is that it's a single wavelength – not a broad band of wavelengths like visible light. Astronomers measuring its intensity and the slight Doppler shifts from the 21.1 centimetre 'rest' wavelength (caused by the motions of gas clouds towards or away from us) can tell not only how much gas there is, but *where* it is in the Galaxy.

Dutch astronomers first predicted that hydrogen should have this telltale signature during the dark days of the Second World War. It was a remarkable achievement. As we've seen in other chapters, astronomers often make new discoveries when they point a new kind of telescope or radiation detector at the sky, and to their amazement pick up strange signals – only later working out what kind of an object they've stumbled across. That's how Jocelyn Bell discovered pulsars, for example, as we saw in Chapter 9. The hydrogen story was rather different.

Back in the 1920s, Dutch astronomer Jan Oort had been one of the first astronomers to produce evidence that our Milky Way Galaxy is a spiral, by studying the motions of stars near the Sun. His pioneering spirit came to the fore again in the 1940s, when the Germans occupied the Netherlands and closed the observatories. Undaunted, Oort insisted on holding his astronomy group together by tackling purely theoretical problems. A few years earlier, American radio engineer Karl Jansky had discovered radio waves coming from the skies – but no astronomers had been interested, because they were absorbed in their own work and didn't want to dabble in radio

12 RIGHT: *Comet West became a brilliant sight in the dawn skies of March 1976, with a glowing fan-shaped tail millions of kilometres long. The broad, curved sweep of the fan is the dust tail, composed of tiny solid grains of rocky dust; the gas tail appears as the narrow straight and bluish 'rib' at the right-hand edge of the fan. At this time, Comet West was moving out from its closest approach to the Sun, and as comet tails always point away from the Sun, the comet was then travelling tail-first.*

13 BELOW: *The Sun appears completely different when 'seen' by the X-rays it emits. This is one of the 150 pictures of the X-ray Sun taken by astronauts on the Skylab space station in 1973 and 1974. The 'surface' of the Sun is not hot enough to produce X-rays and it consequently appears here as a black globe.*

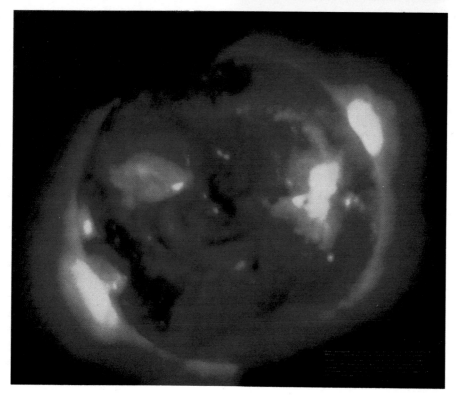

14 ABOVE LEFT: *During a total solar eclipse, the Moon blocks off the Sun's brilliant disc, and the faint pearly corona can be seen shining out around. This eclipse was photographed from the coast of Tanzania in October 1976.*

15 BELOW LEFT: *A cloud of gas and dust near the star rho Ophiuchii is the birthplace of new stars. Portions of the cloud's near-side are lit up by bright stars, but most of the cloud is a dark mass of tiny dust grains which totally absorb any light trying to get through. Astronomers have, however, picked up infrared radiation from a cluster of forty stars within the cloud, stars which have been born from the dense central gas and dust concentration within the past few million years.*

16 BELOW: *A view towards the centre of our Galaxy, in the constellation Sagittarius, shows the thickening bulge of stars around the central regions. The densest concentration of stars, in the Galaxy's very heart, lies in the middle of this picture, but we cannot see it with optical telescopes because the tiny dust grains in space absorb virtually all the light, creating the dark silhouette band seen crossing this photograph.*

17 ABOVE LEFT: *The world's first reusable space vehicle, the American Space Shuttle Columbia, blasts off on its third trip to space on 22 March 1982. The winged Orbiter section carried the two-man crew, the five controlling computers, the main rocket engines and scientific experiments – flown for the first time on this mission. The two small solid-fuel boosters parachute down after launch, and are refurbished for another mission. The only non-reusable part is the large fuel tank strapped to the Orbiter.*

18 BELOW LEFT: *The robot explorer Viking 1 tests the red soil of Mars for signs of life. This panorama was photographed on 17 February 1977, and it shows (lower right) the trenches dug during the previous week by its remotely-controlled sampling arm – the deepest parts go down 30 centimetres (12 inches) into the martian desert. The Viking's meteorology boom (centre) monitored the local weather conditions.*

19 BELOW: *Computer-enhancement has emphasised the subtle colour differences within the 'exploding galaxy' Centaurus A. Lying about 16 million light years away, this giant elliptical galaxy is basically a huge ball of a million million old red stars. The colour-enhanced photograph brings out details of a strange band of gas, dust and young stars which is wrapped around the galaxy; the red and blue patches are regions where stars have formed at different times in the past. X-ray telescopes have 'seen' explosions at the very centre of Centaurus A – possibly from gas surrounding a very massive black hole – and the huge invisible clouds of gas ejected can be detected by radio telescopes.*

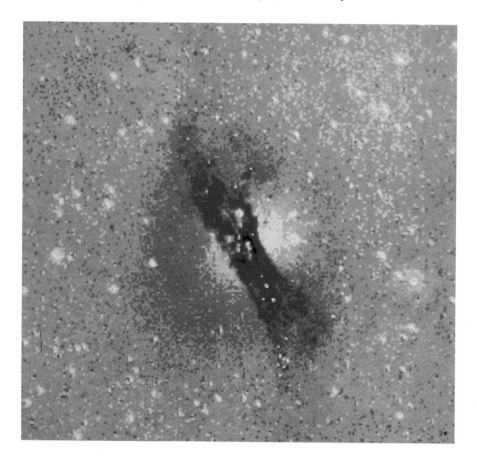

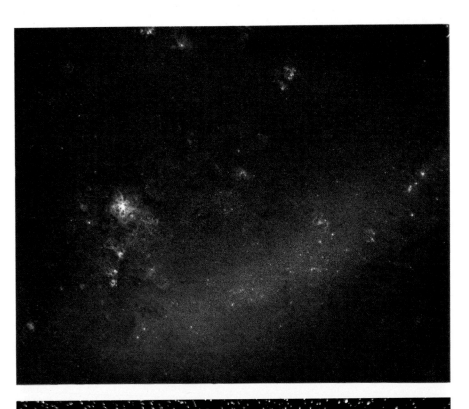

20 ABOVE LEFT: *The Large Magellanic Cloud is the closest galaxy to our Milky Way, lying at a distance of approximately 160,000 light years. It is an irregular galaxy, one-quarter the size of our own, with a wide range of young and old stars, and a great deal of hydrogen gas. Most of this gas is cold, but some has collected together into large hot glowing clouds in which young stars have just been born.*

21 BELOW LEFT: *Visible as a misty oval with the unaided eye, the giant Andromeda Galaxy lies 2¼ million light years away and is the nearest spiral galaxy to us. Although broadly similar to our own Galaxy, the Andromeda Galaxy is half as large again. Its nucleus is made up of old red and yellow stars, while its outer disc and spiral arms are rich in gas, dust and young blue stars. It has two small elliptical companion galaxies, plus a swarm of outlying dwarf galaxies.*

22 BELOW: *The Holmdel, New Jersey, site belonging to the Bell Telephone Laboratories has seen two of the great astronomical discoveries of this century. In 1931, Carl Jansky found radio waves coming from the Milky Way with a primitive aerial at this site. Then, in 1965, the horn-shaped antenna pictured here – designed originally for satellite communications – led Arno Penzias and Robert Wilson to their discovery of the microwave background radiation, believed by most astronomers to be direct evidence for a Big Bang origin of the Universe.*

23 LEFT: *The huge Arecibo radio telescope on the Caribbean island of Puerto Rico consists of a wire mesh dish 305 metres (1000 feet) in diameter, hung in a natural hollow in the hills, with the receiver strung from the surrounding concrete towers. In November 1974, scientists used a powerful transmitter to send a radio beam carrying a message in the direction of a cluster of stars some 24,000 light years away – Man's first intentional message to alien civilisations.*

24 BELOW: *The dense cluster of galaxies Abell 754 is rich in hot gas, which reveals its presence by emitting large quantities of X-rays. In this 'photograph' made by the Earth-orbiting Einstein X-ray Observatory, the gas outshines the many thousands of galaxies in the cluster.*

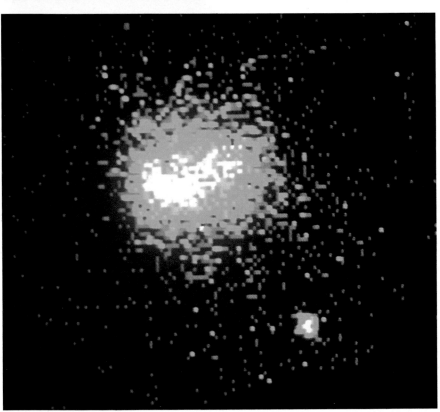

techniques. Oort, however, realised that radio astronomy could extend his own researches into the structure of our Galaxy – if the gas in space should happen to emit radio waves of just one wavelength. He set this problem to a student, Hendrik van de Hulst, who calculated that hydrogen atoms would produce the 21-centimetre waves. In 1951, several radio astronomy groups around the world succeeded in picking up this radiation from the hydrogen in space, and began to lay bare the spiral backbone of our Galaxy.

Oort's change from being an 'optical astronomer' to a 'radio astronomer' marks the beginning of a new spirit in astronomy, a willingness to use new types of 'eyes' to survey the heavens: forty years on, it is as commonplace for astronomers to move from a radio astronomy observatory to a traditional optical observatory, then to an Earth-based centre for orbiting unmanned telescopes picking up X-rays, ultraviolet or gamma rays. And Oort himself is still at the forefront of research. Although now in his eighties, he is still investigating the structure of our Galaxy – and in particular the still-mysterious central regions that we shall look at later in this chapter.

Coming back to the disc of our Galaxy, astronomers have long known that the interstellar gas is piled up into huge clouds hundreds of light years across. These are made dark and brooding by enormous numbers of cosmic dust grains, and nobody knew what might lurk inside – until recently. Radio astronomers have now been able to slice through the dusty veils and have found, to their astonishment, not just hydrogen gas, which is made up of single atoms, but a rich and varied collection of molecules consisting of two, five and even thirteen atoms joined together. To date, over fifty varieties have been discovered. The vast majority are simple organic molecules, which some astronomers claim as support for their theory that life arose initially in space.

Considerations of life apart, the molecules are a fascinating bunch. There are very simple varieties, like carbon monoxide, water and cyanogen, and decidedly complicated eight- and nine-atom compounds such as methyl cyanoacetylene (CH_3C_2CN) and cyanohexatri-yne ($HC_2C_2C_2CN$). There's even one which has been suggested as an incentive to develop interstellar travel: ordinary, perfectly drinkable, but gaseous, ethyl alcohol (C_2H_5OH) – in one cloud complex alone there is enough alcohol to fill the Earth with whisky! The role the molecules play (and indeed, their very existence) is mysterious. They are hopelessly unstable in the hostile environment of space, and so it's not too surprising that they are found only in dark clouds, protected from damaging radiation. But still their lifetimes must be short, which means that the molecules we see at the moment are young, and relatively newly-formed. There must be molecules being born out in space at the moment: the question is, where? The ubiquitous dust grains can provide surfaces upon which the simpler molecules are able to form, but as yet, no agreed birthplaces have been found for the more complex varieties.

The structure of the interstellar medium rivals its composition in being a great deal more complicated than anyone could have predicted. Some astronomers have likened its texture to that of a Gruyère cheese. Seen on a medium scale, it is made up of great arcs and loops of gas which are punctuated by low-density bubbles – evidently, regions swept clean by ancient supernova explosions. But on the very largest scales, the structure becomes simple again. Tuning in once more to the 21 centimetre line from cold hydrogen, we find that – while its distribution isn't uniform – it is spread throughout the Galaxy in a pattern we have met already. Like the young stars and clusters nearer home, it preferentially clumps together along spiral arms. Our Milky Way is, very definitely, a beautiful, slightly open, spiral galaxy.

Proof at last: but with these observations come more problems. The whole Galaxy is rotating, but not like a solid body. Gravity makes its inner regions swirl round far more rapidly than those parts as far out as the Sun, so that a star in an inner orbit will do a lap in only a few thousand years – a trip which takes our Sun 250 million years to complete, even though it is travelling at almost 1 million kilometres (620,000 miles) per hour! The net result of this *differential rotation* should be to wind up the spiral pattern after just a few revolutions, but this clearly hasn't happened. Nor do any of the millions of other spiral galaxies show any signs of being wound up. One of the biggest puzzles facing astronomers today is to explain how spiral galaxies keep their shape.

The field of galaxy dynamics is a horrendously complicated one, in which there are nearly as many theories as theorists. And as leading dynamicist Alar Toomre reflected in 1977, the end is still not in sight: 'the old puzzle of spiral arms ... continues to taunt theorists. The more they manage to unravel it, the more obstinate seems the remaining dynamics.' The unfortunate theorists are aided in their calculations by huge, number-crunching computers which simulate the motions of hundreds of thousands of 'stars' around a 'galaxy'. When it's time to compare theory with reality, the computer galaxies are matched up to the real thing – but there are usually one or two obstinate shapes which refuse to fit.

Nevertheless, astronomers are coming to agree with each other as to what prevents spirals winding up. Best-buy model – although not universally accepted – is C.C. Lin and F.H. Shu's elegant 'density wave' theory, which goes roughly along the following lines. Take a young, rotating disc galaxy which is showing the rudiments of two spiral arms (in case this sounds like cheating, try stirring cream into coffee to see how easy it is to generate a spiral pattern!) Now take a snapshot of the same galaxy a few million years later. In following their own orbits about the galaxy, the previous 'arm stars' have moved on into the general 'inter-arm' region, but their places in the arm have been taken by other stars. The stars in an arm are, therefore, constantly changing. Although this may sound as though the arm will quickly lose its identity, gravity ensures that this is not the case. The very existence of the arm means that gravity is stronger here – because of the bunching of stars – than outside. As stars move around the galaxy, they are actually pulled quickly towards an arm and then, once inside, are slowed down. Getting out of an arm is a slow business too; and then it's a glide between arms before being accelerated on to the next. The net result is that a star spends longer than it should in a spiral arm, and so at any given time, more stars live *in* the arms than between them. The arms are simple density fluctuations in the galaxy's disc where the gravity is higher: perturbations which, once set up, will slowly propagate around the galaxy forever, preserving its spiral shape.

The density wave theory also predicts why the spiral arms of our Galaxy – and of all the other spirals we see – are in such a ferment of activity, busily conceiving and creating new stars. The reason is that the gas in the disc is affected by the gravity fluctuations, too. As the gas is suddenly slowed down on entering an arm, it suffers severe compression, which leads to a rapid burst of star formation. Many of the stars born in this way are massive blue supergiants, whose brief, butterfly lives begin and end in the same spiral arm. As they die spectacularly as supernovae, these stars spew out more star-building materials, and their expanding shock waves squeeze any uncompressed gas in the arm into dramatic collapse. The arms of all spiral galaxies are regions of birth and death, drama and power struggles. They look appropriately spectacular too – their curving outlines are strewn with glowing gas-clouds and sparkling young star-clusters, and their trailing edges are raggedly rimmed by swathes of black, choking dust.

If passage through a spiral arm is a traumatic experience for a gas cloud, how might it affect a star? In particular, what changes might it wreak in the Sun – or even the Earth? Since its birth, the Sun has lapped the Galaxy some twenty times; it has passed through spiral arm regions dozens of times, each passage lasting a few million years. Some astronomers have started to take quite seriously the notion that the extreme conditions in spiral arms could have quite dramatic effects on the Sun and its planets. 'There is an increasing possibility – even likelihood – that external factors may be just as important in an explanation of many of the abrupt changes on the Earth over the past few hundred millions years', claims Sussex University's Bill McCrea.

Although not all the changes can have been triggered by passages through spiral arms, McCrea and his collaborators point to very long-term events which may be a result of changes in the Earth's climate over periods of 100 to 200 million years, roughly the Sun's period around the Galaxy. One case in point is evidence for as many as six mass extinctions of major biological species. Another is the evidence, discussed in Chapter 2, for very long-term climatic variations upon which the shorter period Milankovitch cycle of glaciations is superimposed.

The coincidence in timescales is tempting, but *how* could a spiral arm precipitate sudden changes in the Earth's climate? McCrea's first instinct was that the increased concentration of cosmic dust in an arm might have some effect on the Sun's output, with consequences for the Earth. But only an extraordinarily dense dust cloud could produce any significant blocking in the amount of the Sun's radiation reaching the Earth. However, even an ordinary cloud would prevent the solar wind from reaching us. 'At least one such dusty encounter might be expected on each revolution of the solar system about the galactic centre', says McCrea. However, he isn't yet prepared to predict how the Earth will suffer, but suggests instead that there will be 'possible climatic consequences of an as yet undetermined nature.'

Perhaps more effective, and certainly more dramatic, is the other mechanism proposed by McCrea and his collaborators. For reasons which we have already discussed, the majority of supernovae explode within spiral arms. Although it's difficult – particularly within our own Galaxy – to work out the frequency of these explosions, astronomers observing supernovae in galaxies like our own suggest a rate of two or three per century. The question which all astronomers are asked is, of course: 'What if one went off near the Sun when it was in a spiral arm?'

To do real damage to us, a supernova would have to be closer than thirty light years away (and luckily, there are no supernova candidates within this distance at the moment!) Even then, it would be a very rare event. McCrea works out that it would happen every few hundred million years – but its effects would be disastrous.

Travelling at the speed of light, the short-wavelength radiation from the blast – ultraviolet, X-rays and gamma rays – would take thirty years to reach Earth. Their arrival would be heralded by the utter and instant destruction of our protective ozone layer, allowing these deadly rays to penetrate to ground level and wreak sudden, complete destruction. Hardly anything could survive.

Although the ozone layer could recover in a matter of ten years or so, it would not be given the chance. Streams of slower-moving atomic particles from the supernova would continue to bombard Earth for many centuries afterwards. Although the ozone layer could partly recover, any life-forms struggling to exist would be exposed to much higher doses of solar ultraviolet radiation than usual, and as we have already seen in Chapter 2, severe mutations could result.

The climate would suffer too. The absorption of ultraviolet rays from the Sun by the ozone layer heats the upper atmosphere, and with the ozone layer removed, the temperature would fall. This cooling would be accelerated as a nitrogen dioxide smog closed in around the Earth to replace the missing ozone (radiation would make the nitrogen and the oxygen in the atmosphere react together to form nitrogen dioxide). Soon, the Earth would be likely to find itself in the grip of a major ice age which might last for millions of years.

Astronomers are not doom-mongers, and to be fair, the majority of them think that radiation from a nearby supernova would pose a less severe threat than in the scenario sketched out here. But some claim that an even more serious menace lurks, unseen, 30,000 light years away. It is nothing less than the very heart of our Milky Way Galaxy.

In common with other spirals, the central bulge of our Galaxy is made up of old, red stars which formed long ago, shortly after the collapse of the halo. Unlike the surrounding disc, the bulge has virtually no dust and gas. Until quite recently, it was considered to be an inert, uninteresting place, and its innermost regions could anyway never be seen: the dust-grains in our Galaxy's disc produce more than thirty magnitudes of absorption along the way – that's a dimming factor of thousands of millions of times.

Mapping the spiral structure of our Galaxy with the 21-centimetre hydrogen radiation back in 1957, the pioneering Dutch astronomer Jan Oort was able to penetrate closer to its centre than ever before. He noticed something strange. Instead of spinning uniformly about the centre, some of the spiral arms he dould detect appeared to be moving outwards – away from the centre. Intrigued, he plunged on inwards. He found several more gas clouds, some of them fragments of arms, even jets, which all shared this outward motion.

As the resolution (the ability to see fine detail) of radio telescopes increased in the 1960s, other astronomers reported peculiar features close to the centre of the Galaxy. There is a ring of dark clouds, rich in molecules, measuring 1500 light years across, expanding at the rate of half a million kilometres (310,000 miles) per hour. Embedded in the ring are huge star-forming clouds, each ten times the size of the Orion nebula. Inside the ring – in a part of the Galaxy not famed for its abundance of interstellar gas – is an extended region of hot, excited hydrogen centred on the Galaxy's heart.

There are stars here in mind-boggling quantities. Although we can never see them, infrared astronomers, slicing through the intervening dust, report that the stars appear to be being born within a light year of our galactic centre. Surrounding the centre is a dense star cluster of a million ancient red and yellow stars, so tightly-packed that the distance between near-neighbours dwindles from the four light years in the neighbourhood of the Sun to just four *light days*. And at the heart of all this, there lurks a powerful, but extremely small source of radio waves.

All this youthful, explosive activity points to some kind of disturbance at our Galaxy's heart. As we shall see in Chapter 15, if this *is* the case, and it's a big 'if', our Galaxy is not alone in its affliction, and even so, it is only mildly affected.

But what could be the source of all this unrest at the core of our Galaxy?

'It' is small – less than the span of Jupiter's orbit about the Sun; 'it' is powerful – it can drive away gas streams into space; 'it' gives out energetic radiation. Many astronomers hold the opinion that the core is dominated by a massive black hole – some five million times more massive than the Sun – which was created early on in our Galaxy's history. We will be considering further the question of galaxy formation in Chapter 16; at this stage, it's sufficient to say that a massive black hole may form at the centre of a young galaxy because gas collapses rapidly here.

Black holes, as we have found earlier, are potent engines. Gas and stars which come too close are snatched, torn, swirled, heated, accelerated. The accretion disc – that whirlpool of debris swirling into a black hole – glares fiercely before its matter spirals in. It's this glare which radio (and X-ray) astronomers may be picking up from the direction of our galactic centre.

But accretion discs don't simply glare. Their breakneck spin makes them powerful generators of shockwaves; shockwaves which can drive away nearby gas clouds at explosive speeds. Perhaps we are seeing the results of such shocks in the outward-moving gas streams we monitor today.

Currently, the heart of our Galaxy is not very active. There is probably not enough material in the accretion disc to glare with any conviction, to shock gas clouds into flight or to trigger star formation. But judging from the motion of the outward-moving gas clouds, it has been active in the past – just 13 million years ago, and perhaps only a million years ago, say supporters of the idea. If enough material settled on our galactic core, they claim, it could 'go critical' again, with consequences which could be as devastating for us as the explosion of a nearby supernova. The dire predictions of the doom-mongers may turn out to be wrong, however, for it does seem that we live too far from the centre to be affected.

For the majority of astronomers, these new findings take on a different perspective. It is fascinating, and not a little exciting, to realise that our old Galaxy, for so long regarded as a serene, unchanging whirlpool of stars, has a sting in its tail. And as leading Soviet Academician Iosef Shklovskii put it, somewhat clairvoyantly, as long ago as 1957: 'We have to face the fact that there is some outstanding peculiarity at the galactic centre, whose study will become one of the central problems of astrophysics and cosmology.'

— 12 —

Reach for the Stars!

'Space isn't remote at all. It's only an hour's drive away if your car could go straight upwards.' With this typically striking new perspective, Sir Fred Hoyle emphasises just how near we live to the alien world of space – the dark, empty reaches beyond the Earth's atmosphere. In the bustle of every-day life, many people travel for an hour to get to work each day: if we had the freedom to move upwards as easily as we can travel horizontally, those commuters effectively make the trip to space and back every working day. (There's no real boundary between Earth's layer of air and empty space, in fact, because our atmosphere peters out gradually with height, but scientists take a height of 100 kilometres (62 miles) as a convenient point to say that space begins.)

That comparison is not as bizarre as it may sound. Since civilisation began, Man has been striving to overcome the gravity that holds us to the Earth's surface, and reach upwards. Modern cities bristle with tall tower-blocks, where we may live, or work, or just eat in the top-floor restaurant and enjoy the view. The world's highest structures reach almost one-hundredth the way to space. And today we think nothing of flying in a plane which cruises more than a tenth of the way up to the boundary of space. And although the vacuum of space, and immense temperature changes from sunlight to shade, mean that a man in space has to be well-protected, space itself is no more hostile than the cold and crushing depths of the Earth's sea-beds which man explores by submarine. A well-planned expedition to space is no more dangerous than an exploration of the Antarctic or the tropical rain forests.

Man has already made himself at home in space. Over one hundred people have already been up above the atmosphere, and many have made more than one trip. John Young, commander of America's first re-usable Space Shuttle, was making his fifth journey into space when he took Columbia into orbit in April 1981. While the American missions have carried only United States citizens, the Russians have launched cosmonauts of many nationalities, including a Cuban and a Vietnamese. In the near future, they intend to fly their first non-communist spaceman, a French cosmonaut – 'cosmonaut' being the Russian equivalent of the American 'astronaut'. Between them, the astronauts and cosomonauts have logged more than seven years in space. A significant proportion of this time stems from the marathon efforts of one Soviet cosmonaut, Valeri Ryumin, who has spent in total almost one year of his life in orbit around the Earth. Space travel is now so ordinary that the news media often don't even mention when another Soviet spaceship is launched to dock with their orbiting space station.

But manned spaceflight is only the glittering tip of an enormous iceberg. Non-manned satellites affect the day-to-day lives of every one of us here on the Earth. Satellites transmit telephone calls and television programmes around the world; they help meteorologists give more accurate weather forecasts; they enable ships to find their precise positions at sea; they pinpoint new mineral resources on Earth; and spy satellites have altered the rules of international diplomacy, as they constantly keep their keen eyes trained on unfriendly countries and on potentially-dangerous conflicts.

Many countries now have satellites in orbit, usually launched by the cheap, mass-produced American or Russian rockets. But these two are not the only nations in the 'space club' – the countries which have developed rockets powerful enough to propel satellites up into orbit. As well as the United States and the Soviet Union, the 'club' contains the United Kingdom, France, Japan, China, and, most recently, India. Both Japan and China have launched their own satellites regularly since 1970; India joined the club ten years later with a small satellite called Rohini. Indian prime minister Indira Gandhi said 'India was a late starter, and while others were going ahead we had to keep running just to stay in place.' It is a tremendous task to develop from scratch an entire space programme – rocket, satellite and support systems on the ground. India's example, in particular, demonstrates how most governments now regard an independent 'gateway to space' as vital to their country's interest – in much the same way that governments of a century ago were concerned to have their own strong fleets of merchant and naval ships. The eighth member of the space club is an international venture, the European Space Agency. Its rocket, Ariane, is made up of components from 66 different European companies.

For all their glamour, fire and thunder, rockets are really only a way of getting things higher than planes are able to. An ordinary aeroplane engine burns its fuel in oxygen drawn in from the surrounding air. Such an engine won't work in airless space, and we must use a rocket, which carries not only fuel but also liquid oxygen, or some chemical that can provide oxygen. Once a rocket gets beyond the atmosphere, however, it can do something that no aeroplane can: it can stay up there without any further effort, following a circular or oval *orbit* about the Earth. Scientists call this state 'free fall': the rocket is falling all the time towards the Earth, but it is travelling sideways so fast that its path always misses the Earth. Such orbits are of course common in astronomy. Not only do rockets orbit the Earth, but planets are in orbit about the Sun and the Sun and other stars follow orbits about the centre of the Galaxy.

Most rockets are launched to put a satellite – or several satellites – into orbit. And the total number launched to date is incredible: from the first little Sputnik in 1957 to the end of 1980, Man had launched 2145 satellites. During the 1970s, Russia alone was launching two satellites a week. If we include all the oddments that go into orbit with a satellite, like the final stage of the rocket and the shroud that covered the satellite on launch, then we find that Man has littered space with over 12,000 bits and pieces. Fortunately, they're not all still up there: almost two-thirds have come back down. Satellites in low orbits feel a very slight amount of air resistance from the tenuous outer reaches of our atmosphere, and they gradually spiral inwards until they burn up in the denser lower atmosphere – though some fragments may hit the ground, like portions of the American Skylab space station which burnt up over Perth in 1979, splashing the sky with red and blue sparks, and scattering blobs of metal over the deserts of Western Australia.

Satellites are a vital part of Man's break-out into the third dimension. Some of them have a special reason for being above the atmosphere, such as X-ray astronomy satellites which observe radiation from space that can't penetrate Earth's air and reach ground-based observatories. But the majority of satellites look Earthwards. They simply carry existing techniques to very high altitudes: they are an extension of the activities that we've previously carried out using high towers – in the case of communications satellites – or aeroplanes. From their high station, satellites can survey huge regions of the Earth, and unlike aeroplanes they don't have to expend fuel to stay aloft.

Earth resources satellites, like America's Landsat series, are a very high altitude version of aerial photography. Landsat takes photographs not of individual states or countries, but of the entire world, regardless of national boundaries. Its pictures are taken through different coloured filters, and when scientists analyse the pictures radioed back to Earth they can tell what kind of terrain is exposed. Landsat pictures have tremendous importance in agriculture, particularly to countries where the population is sparsely settled. It's possible to tell one kind of crop from another in the pictures through the use of different coloured filters, and to spot outbreaks of disease. Landsat photographs of the Sahara have shown up regions where the desert has blossomed after a local rainstorm – regions where locusts are likely to breed in huge numbers. And in locations as diverse as Colorado and Libya, Landsat's results are being used to check that irrigation doesn't use up underground water faster than it's replenished, as the pictures show up the effects on surface vegetation.

Where rocks are exposed, Landsat can distinguish one type from another, and so provide clues for geologists seeking new caches of oil and valuable minerals. Many countries now use the freely-available Landsat photographs for this purpose. The Russians have relied on similar photographs taken by their cosmonauts, and have located at least two new oil deposits in the deserts of the Central Soviet Union, unsuspected iron ore in the Ukraine, and valuable ores of the rare metals mercury, antimony and zinc in the Tien Shan mountains.

We're all becoming familiar with the weather satellites' views of Earth, from television weather forecasts. As well as showing the cloud patterns, these satellites can measure temperatures, and they particularly help meteorologists by showing storms that are brewing in remote regions where there are no weather stations. In 1979, a United States government report estimated that weather satellites saved the country about $170 million every year – in applications as diverse as accurately forecasting fruit-killing frosts in Florida, to cutting out unnecessary flights by the coastguard iceberg patrol planes when satellites see no sign of ice. And the savings are not just financial. Satellite pictures can show the precise path of hurricanes and other bad storms. Accurate predictions based on these photographs have already saved countless hundreds of lives in the hurricane belt of the southern United States.

But not all the big eyes in space are so benign. More than half the satellites ever launched have been spy-satellites – the modern equivalent of the post-war 'spy-planes' such as America's infamous U-2. Some of these satellites carry television cameras, and transmit their pictures back to Earth, like the weather and Earth resources satellites. But such messages can be intercepted, and some military satellites instead take photographs on ordinary film. The satellites drop these back to Earth in canisters which are either recovered on the ground, or are caught by high-flying helicopters or planes carrying nets – a deadly serious version of seaside shrimping.

The spy satellites' cameras can show incredibly fine details of the ground. America's Big Bird satellite can distinguish something as small as half a metre (1.6 feet) across; some reports say it can tell military from civilian personnel on the basis of their uniform – and that's from 300 kilometres (186 miles) away, almost the distance of London from Paris!

American and Russian spy satellites not only watch each other's territory; they keep an eye on potential trouble spots. Satellites must be manoeuvred if they are to pass over the same spot on Earth on each orbit; and both American and Russian military satellites have been steered to pass regularly over particular events – including the wars in Bangladesh in 1971 and the Middle East in 1973, and the site of India's first nuclear test in 1974. They were obviously extra-terrestrial spies. Military satellites also eavesdrop on other countries' radio transmissions, and pass messages around the world like ordinary communications satellites. If the United States and the Soviet Union were ever at war, then each would need to destroy the others' satellites. Some satellites are already flying for just this purpose, and Russian killer satellites have successfully destroyed several dummy targets in orbit. It's a sad comment on humanity that war is the most important human activity – in terms of financial outlay – which is being taken up into the dimension of space.

But on the positive side, communications satellites have helped to make the world a smaller and more intimate place. We now expect to see major events, like the Olympic Games, as they happen, wherever in the world they are going on. Most large businesses couldn't work efficiently without easy telephone links around the world, and most of these communications are carried by radio links with satellites. A message sent up from one part of the Earth is picked up and amplified by the satellite, which then re-transmits it to a listening radio 'dish' half way round the world. The latest international communications satellites can handle several television channels and over 10,000 phone calls simultaneously. Despite the high cost of building and launching such a satellite, it can cope with so many messages that the cost per telephone call is only one-tenth of what it would be if the call were made as a result of laying a new ocean cable. About two-thirds of all international phone calls now travel via one satellite or another.

American companies are now using satellites to communicate from one branch to another – often linking one of their computers directly to another. The Mormon Church even intends to beam messages via satellite across the United States to believers who can afford their own receiving aerials. By the end of the 1980s, there should be at least thirty communications satellites serving the United States alone.

It would be very inconvenient if communications satellites orbited the Earth in only 1½ hours, like the Earth surveillance satellites, because Earth-based antennae would then have to track a rapidly moving target. But there's a simple answer, first spotted in 1945 by a young British scientist Arthur C. Clarke – now world famous for his science fiction epics, such as *2001: A Space Odyssey*. The farther a satellite lies from the Earth, the longer it takes to complete one orbit. If it lies 35,800 kilometres (22,200 miles) up – about one-tenth the way to the Moon – it will go round the Earth once in 24 hours. But the Earth itself is turning around, completing one revolution every 24 hours. Thus the solid Earth and distant orbiting satellite revolve together, as if the satellite is on the end of a long spoke. From our viewpoint on Earth, the satellite seems to hang forever over one point on the planet, the point which lies at the foot of the imaginary spoke. Wherever you may be on Earth, the satellite stays in the same point in the sky, and you can keep a fixed transmitting or receiving aerial pointing straight at it.

Almost all communications satellites are in such *geosynchronous* orbits, spread out over the Earth's equator so that they can keep in touch with the whole world. Each of them is effectively a very tall communications tower, like the familiar concrete and iron towers whose small radio dishes beam microwave messages across the countryside. But there's no reason why the satellite shouldn't instead be a super-tall television mast, and broadcast programmes over a wide region where anyone with a suitable aerial can pick up the transmissions. The Canadians have in fact been broadcasting television to their sparsely populated Northern Territories by geosynchronous satellite since 1972 – the satellites are called Anik, Eskimo for 'brother'. In 1975, the Indians launched a satellite whose broadcasts are picked up by 2400 remote villages, each using a small radio dish about 3 metres (10 feet) across.

By the end of this century, communications satellites may be so powerful that we will be picking up all our programmes from space. Instead of bristling with the present type of TV aerial, our homes will support a dish-shaped antenna, the size and shape of an umbrella, with its 'handle' pointing towards a distant satellite. Another possibility could be 'wrist-watch telephones'. A powerful satellite could receive radio signals from a small radio telephone carried on a wrist band, and transmit it to anyone wearing a similar 'telephone'. We could all stay in touch with friends or business contacts whether we are at home or travelling.

In all these ways, space is becoming a part of our everyday lives, creeping in on us quietly and inevitably, much as the electronic microchip has done. But Man is not content to let robots always act for him. Man has always wanted to explore, and the new frontier of space is no exception – Man must reach upwards himself, towards the stars.

The first human to get away from Earth, and circle our planet in orbit was the Russian Yuri Gagarin. His flight in 1961 sparked off the great 'space race' of the 1960s; America's President John F. Kennedy issued the challenge on 25 May 1961, with the words 'I believe that this nation should commit itself to achieving the goal of landing a man on the Moon and returning him safely to Earth.' He promised this historic flight before the decade was out. And, with over six months to spare, Neil Armstrong stepped out on to the Moon's surface in July 1969.

The Russians lost the race to the Moon. They certainly intended to send men around the Moon before the Americans did, and they probably intended to land too. But their programme was crippled in 1966 by the sudden death of its chief, Sergei Korolev, and by the failure of its new super-rocket. This G-1 was designed to pack forty per cent more power than America's Saturn 5 moon-rocket. But on its first test in 1969, the G-1 exploded with the force of a small hydrogen bomb. It killed several generals and leading scientists – and Russia's hopes of sending a man to the Moon.

Since then, the Soviet Union has concentrated on regions of space closer to the Earth, in setting up manned space stations in orbit. It had more disasters in the early 1970s, but towards the end of the decade its programme began to run more smoothly. The space station Salyut 6, launched in 1977, has had cosmonauts on board for over half its time in orbit. They were sent up in pairs in a Soyuz spaceship, for a stay of anything from a few days to six months. Sometimes the 'residents' of Salyut 6 have acted as

hosts to another pair of cosmonauts on a brief visit. And Salyut is continually restocked with fuel and supplies by automatic Progress spacecraft.
Many western experts believe that the Russians will soon launch a larger
space station – perhaps several Salyuts joined together – which will be
permanently manned.

The Soviet Union is making man-in-space an everyday affair, and the
United States is now taking a similar line – a very different policy from the
mammoth national effort that put the first men on the Moon. We saw in
Chapter 3 that the Apollo astronauts returned enough Moon rocks and
other information to reveal most of the Moon's secrets to eager scientists.
But a space effort like that couldn't be kept up. In scale, the Apollo
programme rivalled the wartime Manhattan project to produce the atomic
bomb: at its peak, it used the efforts of 400,000 scientists and technicians,
and its total cost was $25,000 million. After the six Moon landings, the
American space agency NASA had nothing to work with but remnants of
the Apollo programme. America's first and only space station, Skylab, was
a converted rocket case, the empty upper stage of a Saturn 5 that should
have carried a final, cancelled Apollo flight to the Moon. Three manned
crews, launched on a smaller sister of the Saturn 5 moon rocket, spent a
total of six months on Skylab during 1973 and 1974.

Life in a space station is not very different from life in any other isolated,
cramped cabin (Salyut has no more space than a couple of large frame
tents) with one major exception: weightlessness. This is not because space
stations are outside the influence of Earth's gravity. Gravity here is only a
little weaker than it is on the ground. But just as the space station is orbiting
the Earth, in 'free fall', so the men inside are also following the same orbit.
If an astronaut is sitting centrally in his craft, then he'll stay there as he and
the space station follow precisely the same path. There's no tendency for
him to move relative to his craft because the gravitational attraction on
them both is the same, so he can float about inside with no apparent weight.

For the long-stay Russian expeditions, Salyut 6 has been made as comfortable as possible. As well as books, taped music and games, they have
video cassettes – including cartoons and feature films – and a two-way TV
system which lets them see and talk to their family and friends on the
ground. When other cosmonauts visit a long-term crew, they bring up mail
and special food delicacies.

Weightlessness is the main problem with living in space. It means you
can't drink out of ordinary cups – the water just drifts out in a big spherical
blob – nor eat anything which produces crumbs, which can't be caught in
'zero-gravity' conditions. The space station crews have adjusted to living a
weightless life without too much difficulty – but their bodies haven't:
the human body is built to work in conditions where gravity pulls on it.
Half the American Skylab crew members suffered 'space sickness', similar
to travel sickness on Earth. In addition, all men in space have lost a couple

of kilogrammes of water from their bodies; and, for some unknown reason, their bones have lost calcium and become brittle.

Some American experts believe that men cannot stay in weightless conditions for more than nine months before their bones become so brittle they shouldn't return to Earth, because their weakened bones would not be able to support their body weight under normal gravity. But the Russians claim their cosmonauts stay reasonably healthy – on a diet of 3000 calories a day, plenty of exercise on a cycle unit and treadmill, and by wearing a tight 'penguin suit' to prevent blood collecting in the lower halves of their bodies.

Studying the effects of weightlessness is however one of the main economic reasons for going into space. Astronauts and cosmonauts can study how materials and plants and animals behave when there's effectively no gravity, and in these conditions remarkably pure new materials can be produced. Skylab astronauts watched two spiders spin webs, to see how they reacted to weightless conditions – and they coped very well, though the strands were thinner than usual. The Russians on Salyut 6 have produced extremely pure crystals of semiconductors, the material used in electronic 'chips'.

The commercial future of space will depend on processing small amounts of extremely expensive materials: the Russian semiconductors are one useful product, which may give them chips more reliable than any made on Earth; another may be new alloys of metals, which can't be mixed on Earth because the heavier component would sink to the bottom, but which stay perfectly-mixed in weightless conditions and would be returned to earth in a solidified state. Scientists and technologists expect to find more and more uses for space manufacture as we become more at home 'up there'. The Soviet Union certainly intends to follow this road. Veteran cosmonaut Konstantin Feoktistor says 'we no longer doubt the feasibility of space stations for the continuous study of our global resources, the Earth's atmosphere [and] industrial activities in an orbit about our planet.'

The Americans have no plans for a permanent space station, but they are concerned to make space flights a routine business. A conventional rocket is expensive because it's all discarded after one flight. The new American Space Shuttle on the other hand is a reusable space vehicle. Most parts of it are recovered, refurbished and refuelled, and used again. For the price of one moon rocket, American can put up a shuttle thirty times.

The main section is the orbiter, a winged craft about the size of a DC9 airliner. It has three powerful rocket engines, but carries no fuel itself. During launch, the orbiter draws fuel from a large tank strapped below its belly, and the launch is assisted by two separate boosters strapped either side of the fuel tank. The two boosters are recovered after launch, and reused; and the big fuel tank, jettisoned when empty, is the only part which is not used again. After a week in orbit, the orbiter comes back through the atmosphere and flies down to land like an ordinary plane – except the landing must be lined up perfectly first time, for the big ungainly bird has no engines to give a second try.

After years of delay, the first shuttle, Columbia, blasted off to orbit in the spring of 1981. The flight was virtually perfect. The flight directors on the ground enthused 'It was fantastic ... I am ecstatic ... It's absolutely amazing. We have a super vehicle up there.' Commander John Young agreed: 'Hey, this performance is outstanding.' Just over two days later, Young and Robert Crippen brought Columbia back through a fiery re-entry to glide to a spot-on landing in California. America's new space age had arrived.

When the programme is well under way, the United States should be seeing a shuttle launch every week. Its major task will be to take satellites up into space and gently push them away to orbit on their own. The Space Shuttle can also pick up and rescue satellites which have broken down, and since 1980 all American satellites have been equipped with a hook for the shuttle to catch. One-third of the satellites will be military, and the rest commercial or for research – including a huge space telescope. But some of the most interesting results will come from the Spacelab. This laboratory won't be a proper space station, because it remains an integral part of the shuttle and doesn't go into orbit on its own. Its scientists live in the shuttle and work in Spacelab, where they can conduct experiments and test manufacturing techniques in weightless conditions.

The shuttle carries a crew of six, and for the first time ordinary men and women will be able to fly in space, as the four scientific 'mission specialists' who complement the trained astronaut commander and pilot. Twenty years ago, both America and Russia drew their first spacemen from the ranks of strong-willed military test pilots. One of their doctors described their ruthless determination in these terms. 'We're not talking about people like you and me. They're unique before they're even selected as astronauts ... those guys really are interested in themselves.'

But the 'shuttlenauts' *will* be people 'like you and me'. They don't need to be in perfect physical condition, and can wear spectacles if they want to. Six of the first forty-one chosen are women – not Amazons, but simply dedicated career women who have an interest in space. To contrast the difference between today's spacemen and the first astronauts, mission specialist Sally Ride says 'we're not god-like infallible people, but technicians and scientists who will make outer space colonies of the future. I think that's an incredible thing to be part of.'

By the year 2000, then, we may expect to see permanently manned Russian space stations, probably reached by a Soviet version of the shuttle. The American Space Shuttle will be busy launching and retrieving satellites, and it may well be visiting an American/European space station made of several converted Spacelab units. Arthur C. Clarke even predicts that ordinary citizens will be taking trips to the Moon, with fares of less than

£1000 at today's prices. Unmanned satellites will certainly be making our planet an even smaller, more thoroughly known world. And, as mentioned in Chapter 4, there may be a manned expedition on its way to Mars.

It's difficult to see the direction of space travel beyond this. On the one hand, Man wants always to reach upwards, outwards; and on the other, spaceflight is so incredibly expensive that it's difficult to justify unless there's a financial reward at the end. There may be profit in space. The Moon is rich in some elements rare on Earth – for example, the tough metal titanium. Some scientists suggest flinging huge amounts of Moon rock towards Earth by some kind of magnetic sled, and processing them in enormous factories orbiting the Earth.

Asteroids could be an even better site for space mining. Nature has started processing their material already: as we saw in Chapter 6, some are almost pure metal, an iron-nickel alloy, while others are rich in carbon, the basis for plastics. It may not even be worth returning these materials to Earth. Future space generations could use them to build huge space constructions which would pay back their own cost – for example, satellites to convert sunlight to electricity.

In the late 1970s, a number of reports looked seriously at the possibility of building these enormous solar-power satellites. Each satellite would be 10 kilometres (6 miles) long and 5 kilometres (3 miles) broad – the size of Manhattan – and would be covered with solar cells which convert sunlight to electricity. This area would produce 5000 megawatts of electricity, about enough to supply an average city. The satellite would beam the energy back to Earth as microwaves, to a receiving station covering some hundred square kilometres (38 square miles) of ground. In 1981, one report concluded that it would cost $3,000,000 million to construct sixty such satellites over a fifty-year period, if the materials were flown up from Earth; but it found no impossible technical problems and concluded that they 'could become an interesting option at some time in the 21st century'. By then solar power satellites made from moon or asteroid material may be a financial success.

Another possibility is to construct huge 'space colonies'. American scientist Gerard O'Neill believes that man will one day escape from Earth: 'unless the whole of the human race manages to blow itself up first, I think it's almost inevitable.' Rather than try to colonise the other planets, he thinks it makes more sense to construct artificial colonies in space. The first would be hollow spheres, each about half a kilometre (a third of a mile) across, filled with air and housing 10,000 people. Each would spin round twice a minute, and centrifugal force would create an artificial gravity acting outwards around its 'equator' – so people standing firmly on the inside would look 'upwards' to see the heads of others standing apparently upside down on the opposite side of the sphere! Larger colonies might hold a million people, and be shaped like huge hollow cylinders.

Biologists, however, have strong doubts about O'Neill's colonies. The Earth's land, sea, atmosphere and plant and animal life form a delicate balance, as we saw in Chapter 2. We probably couldn't recreate this balance artificially, on a small scale. But if it were possible, O'Neill's colonies would become a life raft for the human race. Even if mankind on Earth were destroyed, the population of the colony could survive to perpetuate our species. And O'Neill expects that some colonies would equip themselves with slow but long-lasting rocket engines, and propel themselves ponderously out of our solar system. 'If I had to bet', he says, 'I would bet that a century or so from now there will be a large space habitat on its way to another star.'

O'Neill's colleague Freeman Dyson has proposed an even more audacious engineering feat to increase our living space. Dyson says that we could dismantle a planet, and with its remains build a huge hollow sphere right round the Sun, at about the Earth's distance out. We could live on the inside of this sphere, because each part would be like the Earth's surface at present – the difference being that the whole surface is 550 million times bigger.

If we could build a Dyson sphere, we would be able to trap and use all of the Sun's energy. Russian scientist Nikolai Kardashev has proposed that this would make a second stage in civilisation. His Type I civilisation uses the energy of a whole planet – we're almost at this stage now – while his Type II controls the energy of an entire star. If Man ever makes the grade to Type II, we should also have the technology to reach the stars.

In fact, Man has already sent four unmanned messengers on their way to the stars. The spaceprobes Pioneer 10 and 11, and Voyager 1 and 2, are travelling outwards so fast that they will break free from the Sun's gravity. Eventually they will travel freely amongst the stars. But even though they travel faster than a rifle bullet, such is the immensity of space that a Pioneer or a Voyager would take 100,000 years to reach the nearest star.

The sheer scale of space is the problem that dogs all attempts to talk about interstellar travel. One of O'Neill's giant space arks could hardly travel faster than the speedy little Voyagers, and 3000 generations would pass on board before it reached its journey's end.

We may, one day, be able to put on a faster turn of speed. A group of scientists and engineers belonging to the British Interplanetary Society have designed an unmanned space probe that could reach one of the nearest stars, Barnard's Star, in only fifty years. The probe, Daedelus, is blasted out into space by a continual stream of tiny nuclear explosions behind it, which take the place of the chemical explosion in the combustion chamber of an ordinary rocket. We could build Daedelus today if any government had the money to construct it, and to 'quarry' its heavy hydrogen fuel from the atmosphere of Jupiter.

The Daedelus probe would only fly past Barnard's Star, and then coast on to empty space with no further use. Other scientists have proposed that the first star-probe should be able to slow down and explore the star system it meets; to seek out asteroids automatically; mine them, process the materials, and construct new versions of itself. It would be a self-reproducing star-probe. Its daughters would set out to explore more star systems – at no extra cost to us at home. As each arrived, it would spawn again, until every star in the Galaxy had received its star-probe messenger, each sending information back to us here on Earth.

The original idea was conceived by American computer pioneer John von Neumann, and such self-reproducing probes are usually called von Neumann probes. American mathematician Frank Tipler estimates that, in a century's time, we will have the technology to construct a von Neumann probe for less than the cost of the American Apollo Moon programme. And this one seed could spawn new probes to lay open the whole Galaxy to our automatic messengers.

But we don't know how Man himself will ever cross the yawning gulfs between the stars. Unless we are prepared to let generations pass *en route*, we must find ways of travelling much faster than we can conceive at present. And this means finding engines of tremendous sustained power to boost us up to speeds near to the speed of light – Nature's own speed limit. We can never travel faster than light itself, so our journey to the nearest star, 4.3 light years away, can never take less than 4.3 years. And a trip right across our 100,000-light-year-wide Galaxy must always take at least that number of years.

But we do gain an advantage when moving at these speeds – the apparently strange 'twins paradox'. If one of a pair of twins sets off at almost the speed of light to visit a star ten light years away, his brother on Earth will be slightly more than twenty years older when the astronaut returns. But the astronaut himself will have aged only a month or two. The faster he travels, the slower time passes for him. In theory a space traveller could explore our Galaxy from end to end within his lifetime – as long as he keeps moving fast enough. But when he returns home, his friends will be 200,000 years dead.

Even today, Man can reach the stars in theory, although the reality may be centuries away. But the nearer reaches of the Universe are now firmly part of the human empire, an upward storey to our living space. We are making ourselves at home there, colonising it for our own ends. One American scientist has said 'Apollo was our Christopher Columbus; the space shuttle is our *Mayflower*.' And from the other side of the world, a Russian report acknowledges that their future space missions may include the conception and birth of a baby in space: 'the tempo of space research is so quick that ... the day will come when the first citizen of the Earth will have written on his birth certificate: place of birth – Space.'

— 13 —

Alien Life

In late 1974 American scientists broadcast a unique radio message. The message itself wasn't so unusual, for it merely described human beings, the biochemistry that runs our bodies, and the solar system. But its intended audience was. For the first time, Man was broadcasting not to other humans on this little planet of ours, but to the ears of alien beings on other worlds far out across the Galaxy. The message is a greeting; an invitation for an interstellar penfriend; and an application to join the 'Galactic Club' of alien civilisations which may already be speaking to one another across the length and breadth of our Galaxy.

The message was sent on a powerful blast of radio waves from the world's largest radio telescope, at Arecibo in Puerto Rico, used in reverse as a transmitter rather than a receiver. It is now winging its way outwards at the speed of light. Already, Man's message has travelled further than the distance of the nearest three star systems; by mid-1983 it will be farther off than Sirius, the brightest star in our skies. But the message is heading for none of these stars. The giant radio telescope has beamed it towards a distant cluster of a million stars, M13 in the constellation Hercules, where it will arrive in some 24,000 years time. If anyone there picks up this fleeting broadcast from planet Earth, and sends us a reply, we will receive their response about the year AD 50000.

This broadcast was a brave and optimistic act, for we don't have any evidence that there are civilised aliens living on planets of the stars of M13 – or anywhere else in the Universe, for that matter, apart from planet Earth. But many scientists believe that the Earth cannot be unique. Our very ordinary planet, circling a very ordinary star, has managed to spawn an intelligent race of beings. It would surely be very conceited of us to think that ours is the only planet in the huge cosmos where this has happened.

We don't have definite proof that aliens exist until we pick up an intelligent message from space, or stumble across an unearthly artefact, or meet 'them' face to face. But astronomers are now beginning to know enough about the Universe to estimate just how common, or rare, planets like the Earth should be, and biologists can estimate the chances that living things could have arisen there. As a result, the last thirty years has seen scientists take up 'alien life' as a serious study – even though they can't be certain that the subject of their study actually exists!

There's only one place where scientists can study life itself, and that's on planet Earth. When we think of living things in everyday life, it's generally the larger varieties that spring to mind – humans, dogs, flowers, trees – and they all seem remarkably different. But such living things are basically huge collections of microscopic *cells*, arranged in different ways, and all cells, whether plant or animal, are very similar to each other. The first living things on Earth were single cells, and these later evolved to produce the creatures we know today.

So the origin of life on Earth really boils down to the formation of the first living cells from the inorganic rocks and gases of the early Earth. In those days, the Earth's atmosphere was very different from today's air: as we saw in Chapter 2, it consisted of carbon dioxide, nitrogen and water vapour, with a whiff of the unpleasant gases carbon monoxide, hydrogen sulphide, ammonia and hydrogen cyanide. Such an atmosphere would poison us human beings, but these simple gas molecules were the basic building blocks of the giant molecules which make up living cells. All it needed was some force to weld them together.

And the young Earth's atmosphere was certainly exposed to some powerful forces. Huge meteorites crashed through it, to the accompaniment of immense sonic booms. Unprotected then by an ozone layer, our planet's air was irradiated by searing ultraviolet radiation from the Sun. And the turbulent clouds were rent by vicious lightning bolts. These forces broke up the simple gas molecules, and welded them back together in larger groups, like the amino acids which later strung together to make proteins, and the bases which were to form the core of DNA, the enormous molecule which passes on hereditary characteristics.

Although this phase of Earth's past is long since gone, scientists have confirmed that these processes do work, both by laboratory experiments and from Voyager's look at a natural deep-freeze far out in the solar system. Many chemists around the world have recreated Earth's early atmosphere by mixing the appropriate gases in a glass flask, and disrupting the mixture with shock waves, ultraviolet light or electric sparks to simulate meteorite impacts, solar radiation or lightning. And in every case these forces have indeed welded the simple gas molecules together. The result is a thick 'gunge' of organic molecules – including the amino acids and bases so essential to life.

Astronomers found that Nature does indeed follow the course of these experiments on a planetary scale when Voyager 1 scanned Saturn's large moon Titan in 1980. Titan has a thick atmosphere of nitrogen, contaminated with simple gas molecules like those of the early Earth – in particular, methane and hydrogen cyanide. And these gases have indeed reacted together to make a dense orange smog of organic gunge – the 'clouds' which shroud Titan's surface from view (Chapter 5). American scientist Rudolf Hamel explains that on Titan 'we may have a snapshot of atmospheric evolution which might have taken place a long time ago here on Earth ... we may find in a "deep freeze" the [organic] molecules which may have evolved on [the primitive] Earth.'

On our planet, these molecules must have dissolved in the early oceans. Somehow, the amino acids must have strung themselves together by the hundred to make proteins; while the bases joined up in huge DNA structures with sugar molecules from the organic gunge and phosphate molecules dissolved from rocks. In the process, these super-molecules began to work together. Eventually, bound inside a little protein bag, they became the first living cell, able to grow and reproduce itself.

Over a century ago the great biologist Charles Darwin suspected that life on Earth began something like this. Well before anyone knew how the Earth was born, or had tried experiments to produce the first molecules of life, Darwin conceived of 'some warm little pond, with all sorts of ammonia and phosphoric salts, light, heat, electricity, etc., present', where 'a protein compound was chemically formed ready to undergo still more complex changes' which would produce a living organism. But although we now know that Earth must have been rich in both amino acids and bases, scientists today still aren't sure how exactly they came together to make the first living cells. And, most crucial of all, they don't know if it was inevitable.

Some scientists argue that the simple chemicals of life – amino acids, bases and sugars – are tailor-made to fit together as proteins and DNA. So they would have built up living cells as surely and inevitably as microscopic ice crystals in a winter's cloud build up into beautiful hexagonal snow flakes. These scientists point out that the first cells formed pretty soon after the Earth's own birth. The oldest known fossils of cells, found in north-western Australia, date back 3500 million years. American palaeontologist J. William Schopt says that the Earth was by then 'populated by numerous types of relatively advanced bacterium-like forms of life'. The first primitive cells probably formed at least 500 million years earlier, during the first few hundred million years of the Earth's life – and calculations shown that this is far too short a period for the complex chemicals to have come together by trial-and-error.

On the other hand, unlikely coincidences can happen. Even though the odds are strongly stacked against it, a series of chance meetings of the right molecules could have made a living cell more-or-less by accident. This one cell would then be the ancestor of all cells in all creatures and plants alive today. There is a suggestion that such an unlikely coincidence did occur, in the make-up of modern cells. Some molecules important to life can occur in two different 'mirror-image' forms, and if life had arisen inevitably there should have been a large number of original cells, some with each type of molecule. From these ancestor cells would have descended living cells with both kinds of molecule. But all living cells today contain just one type of these asymmetric molecules, indicating that they have all descended from just one single ancestor cell.

As far as the Earth is concerned, this debate is something of a technicality – whether or not life is 'inevitable', it certainly has happened here. But once we start to calculate whether there's alien life in the Galaxy, then it's crucial. Astronomers have calculated there are something like 100 million Earth-like planets circling Sun-like stars in our Galaxy. Their primitive atmospheres must have produced amino acids and bases, which rain has washed down into their seas. If living cells only arise as an almost impossible fluke, then the Earth is likely to be the only life-bearing planet amongst these; but if the formation of living cells is inevitable in such conditions then all these planets will bear life. In the first place we are alone in our Galaxy; but in the second we could be surrounded by some 100 million different species of intelligent alien life.

Science fiction authors have been inventing alien 'bug-eyed monsters' for at least a century, but scientists only began to take the idea seriously in the early 1950s. One big stimulus was the first 'origin of life' experiment. American chemists Harold C. Urey and Stanley Miller passed a spark through a flask of 'primitive atmosphere', and were surprised to end up with a rich mixture of the simple chemicals of life, including some amino acids. At the same time, astronomers were coming to accept that planets condense from dusty discs around new-born stars (Chapter 6), in place of an older theory which said that the Sun's family of planets was the result of a freak accident. Instead of thinking of planets as a rarity, astronomers started to believe that most stars must have accompanying planets.

If life, or at least its first stages, can arise as easily as Urey and Miller had demonstrated, and planets like Earth are two-a-penny, then there must be aliens out there! Swayed by the seductive argument, astronomers began to think seriously about contacting them across the interstellar void. The science of 'alien life' became respectable in 1959, when the learned journal *Nature* published a letter from American physicists Giuseppe Cocconi and Philip Morrison suggesting that the wavelength of natural radio emissions from the hydrogen atom, 21 centimetres (8.4 inches), would be a suitable wavelength for interstellar broadcasts. Within a year, American radio astronomer Frank Drake was 'listening in' to two neighbouring stars in the hope of picking up a communication from an alien intelligence. Drake called his attempt Project Ozma, after the princess of the land of Oz, a country 'populated by strange and exotic beings'.

Drake had no success – but then it was extremely unlikely that two stars so close should have planets with intelligent, technological civilisations at the same time as ours. Even the most optimistic astronomers believe we shall have to listen in to 100,000 stars before we pick up the first interstellar message. Nothing daunted, radio astronomers around the world have taken up the challenge. The American National Radio Astronomy Observatory and the Canadian Algonquin radio observatory spent years surveying hundreds of stars during the 1970s, while the Russians plan to follow up earlier programmes with a fully automated set of radio telescopes on the ground linked to another in orbit.

The huge radio telescope at the Ohio State University has been listening-in to the sky continuously since 1973. Its director, Robert Dixon, says it could pick up signals broadcast by a similar telescope (used in reverse) up to a distance of 1000 light years. Like all the other listening-in projects, Dixon's marathon effort hasn't discovered a single message from beyond. But he intends to keep on searching, because a message from another civilisation would be the most important communication that Man has ever received. Twenty years after his epoch-making paper in *Nature*, Philip Morrison is still as keen as ever. 'We must make an experimental attempt to find out if there is any signal, any beacon directed at us, or directed generally, to open a communication channel. We should carry on until we can say "we have searched well, and we are alone – probably". I don't think we should stop now.'

Even while these radio astronomers were setting up their huge ears on space, other astronomers were finding clues to bolster their optimism about the possibility of life elsewhere in space. American astronomer Peter van de Kamp was analysing in detail the positions of nearby stars, and concluded that some were being perturbed by planets orbiting them – just as the Sun moves around its mutual balance point with Jupiter (Chapter 5). This was experimental evidence, albeit indirect, that stars in general do have planets. Van de Kamp concluded that the second nearest star, Barnard's Star, has two planets, very similar in mass to Jupiter and Saturn. Odds on, it would have smaller planets like the Earth too.

Astronomers also found that simple organic molecules (precursors of life) are amazingly common in the Universe. Radio astronomers detecting natural radiations from space discovered that dense clouds of gas and dust are thick with simple organic molecules, formed by chemical reactions in space (Chapter 11). Scientists investigating carbon-rich meteorites had another surprise in store. When they analysed a meteorite which fell over Murchison, Australia, in 1969, they found it contained not just simple carbon compounds, but even amino acids (the components of proteins) which had evidently formed by chemical reactions in the meteorite's parent body out in the asteroid belt.

Faced with the evidence for organic molecules in space, Sir Fred Hoyle has taken a typically original viewpoint. With colleague Chandra Wickramasinghe, he proposes that life actually begins in space, not on the surface of planets like Earth. Some of these free-floating bacteria and viruses landed on Earth just after its birth, and gave rise to life here: some still drift in from time to time, and cause sudden epidemics of 'flu. But most are out in interstellar space, in the form of the 'dust' which obscures our view of distant stars. Hoyle and Wickramasinghe believe that 'microbiology operates on a cosmic scale ... interstellar dust grains must all start life as viable bacterial cells ... [but] ultraviolet light would slowly but steadily degrade the material ... eventually to leave graphite particles.'

Most astronomers and biologists have found too many technical problems with this theory to take it seriously. But it is an expression – if an extreme one – of the feeling of optimism in the early 1970s, the feeling that life in the Universe is common, the feeling that we should strive to join the Galactic Club.

And so American astronomers sent that first message from the Arecibo radio telescope towards the star cluster M13. There's very little chance that anyone there – even if they exist – will be pointing a radio telescope in the right direction at the right time to pick it up. But it is a symbol. In similar vein, scientists have attached messages to the four interplanetary space probes which are travelling fast enough to leave the solar system, and will end up journeying endlessly amongst the stars of our Galaxy.

Each of the Pioneer 10 and 11 probes carries a flat plaque, inscribed with line drawings. A simple map shows the location of the Sun, relative to Nature's own lighthouse beacons, the pulsars; another diagram delineates the planets of the solar system; a third feature is a line drawing of two naked human beings, with a hand raised in greeting.

The later Voyager 1 and 2 probes each bear a more sophisticated message – in the shape of a long-playing gramophone record. For maximum durability, the discs are made of copper, and they, along with a suitable stylus and cartridge, are sealed into an aluminium container. The record starts with 115 pictures, encoded as sound signals, of the solar system, the Earth, animals and plants, human beings from all around the world, and human activities and creations – including a string quartet and a supermarket. But any alien who finds a Voyager adrift in space will look in vain for a true representation of the beings which sent it. America's publicity-conscious space agency NASA had some letters of complaint that with the Pioneer plaque it had been 'using the taxpayer's money to send smut into space', so it vetoed naked human beings on the Voyager discs. Following messages from President Jimmy Carter and Kurt Waldheim, Secretary-General of the United Nations, the disc intones greetings in 55 languages. It ends with an 'essay in sound' depicting typical sounds of Earth, and a selection of music ranging from a pygmy girls' initiation song, through Beethoven's fifth symphony, to Chuck Berry's 'Johnny B. Goode'!

None of these probes is heading for any particular star, and there's virtually no chance that a Pioneer or Voyager would be spotted accidentally in interstellar space by an alien spaceship. Like the Arecibo radio message, they are as much an indirect message to people on Earth: the publicity surrounding the plaques and discs has spread the word that scientists believe that there are other civilisations 'out there'; that we should think of mankind more humbly, not as the pinnacle of creation but as just one of many brothers across the cosmos.

In the early 1970s, many scientists took their enthusiastic belief in 'alien life' to extremes, asserting that many interstellar civilisations must be 'so close together that they can't help but discover one another', and that it's possible by now that 'thousands or even millions of cultures may be linked together'. These conclusions are the end result of calculations based on a string of plausible estimates – of how many stars have Earth-like planets, of the chance that life would arise there, of the likelihood that intelligent life would evolve. But such calculations are long on assumptions, and very short on facts. By the mid-1970s, scientists saw that they could check one major assumption at a stroke, by testing another world for life – the planet Mars. If we find living cells on Mars, the first suitable planet we check out, then life must indeed be commonplace throughout the Galaxy.

At first sight, the red deserts of Mars, frozen under a thin dry atmosphere, do not seem a likely habitat for frail living cells. But in 1971 the Mariner 9 spacecraft had photographed dried-up water channels, and this meant that Mars must once have had streams and seas of liquid water, and a thicker, moist atmosphere. If life *can* evolve on any suitable world, then we might expect to find small bacterial cells lying dormant in the red soil of Mars. The two American Viking probes which went to Mars in 1976 carried experiments designed to search for living cells.

These spacecraft carried Man's senses to another world. The two cameras on each lander gave stereoscopic vision; the long finger of a meteorological boom told of the wind direction and speed; a complicated nose (the mass spectrometer) sniffed out and analysed the gases in Mars's atmosphere; and a long mechanical arm equipped with a scoop could dig out handfuls of the martian soil. But for all their sophistication, the Viking landers could not analyse the soil as thoroughly for signs of life as could a biologist in a fully-equipped laboratory on Earth. The 'laboratory' space available on board was only the size of a large wastepaper-basket, and this had to hold the equipment for three different experiments.

The experiments to test for life were thus made as simple and reliable as possible. They couldn't find individual microscopic living cells, so they looked instead for the kind of activity that living cells produce. The first experiment involved pouring a nutrient broth (simple organic 'food' dissolved in water) on to a thimbleful of martian soil. Living cells of any kind should have converted some of this food to waste gases and a detector kept sniffing at the soil to find out what gases came off. And the first result astonished scientists on Earth, because the experiment succeeded too well! A huge amount of carbon dioxide and oxygen bubbled up almost immediately, then gradually died down. But this was the opposite of what scientists expected. Living cells should have produced only a little gas at first, and then gradually more, as they reproduced and grew in numbers. The Viking scientists concluded that they had instead set off a chemical reaction. The martian soil is not inert, like the sands of the Sahara, but contains unstable chemicals which had reacted with the nutrient to produce gases.

These chemical reactions cropped up in the second experiment, which again involved adding a nutrient solution. This time, the experiments wanted to test for 'animal-like' bacteria – cells which take in nutrient 'food', and exhale carbon dioxide, just as we do. Again, there was a surge of carbon dioxide. But once more, the gas could have come from a chemical reaction between the martian soil and the nutrients.

The third experiment was tailored to look for 'plant-like' cells, which absorb carbon dioxide from the atmosphere and convert it into part of their structure. Once again, the experiment was apparently a success. Carbon dioxide from an artificial atmosphere in the test chamber was indeed absorbed into the sample of soil, as if there were active cells there. But knowing that the soil contains reactive chemicals, scientists had to be cautious. They tried heating another sample of soil to a temperature that would have killed any living cells – a kind of pasteurisation – before repeating the experiment. But again they got a positive result. Chemistry must be responsible.

At first, scientists were simply confused by the results pouring in. Viking mission director Tom Young conceded that when the experiments were designed, 'we did not properly comprehend how complex the martian problem was.' Even a couple of years later, two Viking researchers maintained 'on the basis of strict interpretation of the scientific evidence, the possibility that we have detected life on Mars remains a real one.' But most scientists now agree that Mars is a dead world.

The conclusive evidence has come, ironically enough, from a completely different experiment. This simply 'cooked' lumps of martian soil, and sniffed the gases coming off with the same spectrometer used for analysing the martian 'air'. When cooked any living cells on Mars should emit a characteristic odour of organic molecules, like a small-scale version of the Sunday joint. But this incredibly sensitive instrument found no sign at all of organic odours, no indication that Mars's soil contains any carbon compounds – let alone anything as carbon-rich as living cells. Viking investigator Norman Horowitz sums up the generally accepted conclusion when he bluntly states 'at least those areas on Mars examined by the two [Viking] spacecraft are not habitats of life.'

To the one fact we already knew – life exists on the Earth – we can now add another. Life does not exist on Mars. This result doesn't mean that life *can't* exist on planets of other stars, but it did mark a turning point in the scientists' attitude. In the heady optimism of the early 1970s, belief in widespread alien life became almost a religious dogma. It was fashionable to talk about life on other worlds, even though the evidence was meagre and highly circumstantial. In the later 1970s and 1980s, the pendulum has begun to swing the other way. In part, this is simply a natural reaction to the earlier over-optimism. But there's another reason too. As scientists have slowly begun to assemble a few hard facts, they have all turned out to be disappointing. None of these facts actually rules out life elsewhere in the Universe, but they seem to be straws in the wind.

The Viking's failure to find life on Mars was the first disappointment; another has been the radio astronomers' long and fruitless search for intelligent messages from other civilisations, and Peter van de Kamp's 'evidence' for other planetary systems has turned out to be highly suspect. Other astronomers have re-analysed his data, and conclude that the supposed movements of his candidate stars are probably caused by slight distortions of the telescope he uses. Astronomers at another American observatory have measured Barnard's Star – van de Kamp's best case for a planetary system – with more modern and accurate equipment, and they find no sign of all of the supposed movements caused by planets. So there is still no proof that any stars, apart from the Sun, have planets in orbit about them.

Other astronomers have argued that the Earth and Sun are not that 'ordinary'. Scientists once believed that a planet lying anywhere between the orbits of Venus and Mars would have liquid water, and so be suitable for life. There's obviously a good chance that a planet could lie in such a wide ecosphere (region of life) around any star. But later calculations show that the ecosphere is much narrower. If the Earth were only a few per cent closer to the Sun, it would have gone the way of Venus, baked red-hot under a thick blanket of atmosphere; only one per cent farther away, on the other hand, and it would be in the grip of a permanent ice age, like Mars. There's only a small chance that another star would have a planet within such a narrow habitable region, even if it has a planetary family.

And a more detailed study of the other stars shows that the Sun is an unusually sedate star. Although it is fairly average in its output of light, the Sun has less magnetic field activity, and a weaker 'wind' of particles streaming off it. Life would find survival more difficult on a planet circling in the raging wind of particles from a more typical star.

There's one more, very basic fact in the debate on alien life – so obvious that it almost escapes our attention. There are no intelligent alien beings on the Earth; and there are no alien artefacts to indicate that extraterrestrial beings have ever visited our planet, or have sent automatic spaceprobes to investigate our world.

Some authors, such as Erich von Däniken, have in fact claimed that alien spacemen visited Earth in the past, helped our ancestors to erect great constructions like the Pyramids of Egypt, and have been preserved in myths as 'gods'. But their supposed 'evidence' disappears on closer examination.

One person who's tried to analyse such claims seriously is Scottish writer Duncan Lunan. Referring to the evidence left standing after only a 'preliminary analysis', Lunan says 'it carried so little weight that it had to be presented in humorous terms.' American astronomer Ron Bracewell has investigated some of von Däniken's cases in detail, and finds they are based on mistaken 'facts' and misinterpretations – in one instance, von Däniken did not even visit a South American cave system which he later describes 'first hand' in one of his books. Bracewell describes von Däniken's works as mere 'romanticist's fiction'.

On the other hand, many respected people believe that alien spacecraft are flying around our world today, in the shape of UFOs – unidentified flying objects, or flying saucers. They include former US President Jimmy Carter, some of the American astronauts, and American astronomer J. Allen Hynek – a sceptic about UFOs until he was asked to analyse UFO reports submitted to the US Air Force. Despite years of investigation by scientists like Hynek, however, there is still no clear idea of what a UFO is, nor even any evidence that they actually exist.

It was clear that many people have seen things in the sky that they cannot identify. A Gallup poll in 1966 suggested that five million Americans had seen a UFO of some kind, and half a million of these had reported their sightings to the authorities. But the vast majority of these objects are undoubtedly natural or man-made, rather than alien. If a trained investigator interviews the witness, he can usually identify what the 'flying object' was. The bright planet Venus is one frequent culprit, and bright shooting stars another. Daytime UFOs include planes and weather balloons. But investigators like Hynek have found that they can't explain a few per cent of the UFOs – often objects seen independently by several reliable witnesses. Hynek has said that UFOs point to 'an aspect or domain of the natural world not yet explored by science'.

But despite years of serious investigation, there's no evidence to indicate that UFOs are alien spacecraft. In fact, the available evidence points the other way. Almost all the photographs of metallic discs seen during daytime have been exposed as fakes. The most famous case of people (the case of the American couple Betty and Barney Hill) being abducted by a flying saucer, has turned out to consist of imaginative details added afterwards – with a recall, under hypnosis, of a vivid dream, not of the supposed event itself. The notorious Socorro flying saucer, whose inhabitants were seen by a policeman, was only a publicity stunt by the local mayor. And the much publicised UFO seen by Jimmy Carter was nothing more than the planet Venus!

UFOs are not alien spaceships. Whether they exist at all is another matter. Many serious investigators believe that all UFO sightings could be explained away given a sufficiently detailed investigation. American UFO investigator James Oberg points out that the police regularly deal with cases of unsolved crimes, unfound missing persons and unexplained car accidents – but they don't invoke 'some extraordinary criminals, some extraordinary kidnappers, or some extraordinary traffic saboteurs'. Other investigators suspect UFOs may be a psychical phenomenon, akin to poltergeist. Psychical researcher Hilary Evans suggests that even if UFOs don't exist at all, then they still constitute a problem – to sociologists. 'No other case exists in which some half million persons have recorded ... their alleged sighting of an object which is claimed by others not to exist at all.'

If extraterrestrial civilisations are as common as scientists believed ten years ago, then it's very strange that they haven't colonised the Earth, or even visited it. If there are millions of other civilisations in our Galaxy, many of them must be much more advanced than we are. Our Sun is merely a youngster on the astronomical timescale. Some stars are three times older, and any life on their planets should have a head start on us of some 10,000 million years. Although we have just started on the road to space – it's only 25 years since the first tiny Sputnik went into orbit – we can already envisage starships. As outlined in the last chapter, engineers have suggested three different ways to reach the stars using technology available within the next fifty years: slow, ponderous space arks; nuclear-powered 'Daedelus' craft; and the unmanned self-replicating von Neumann probes. Any civilisation which is thousands of millions years ahead of us must have the ability to cruise the stars.

But would they want to? Again, it's a matter of numbers. Some civilisations may be content to stay at home and manage their own small planet. But we know that man has a restless instinct to explore, an instinct which will one day drive our descendants or their robots to the stars, as surely as it has already taken men to the Moon and robot spacecraft to the corners of the solar system. It would be strange indeed if none of these other millions of civilisations had an urge to explore. And the activities of only one such civilisation could be far-reaching. Fanning out from one planetary system to several others around, colonies of beings (or self-replicating robots) could spread across the entire Galaxy and populate every available planet in less than 1000 million years. Since some civilisations should be 10,000 million years ahead of us, they've had plenty of time to colonise the Galaxy.

These arguments no more 'prove' that alien beings don't exist than the optimistic lines of thought a decade ago 'proved' that they do. In truth, we need to know much, much more about the Universe and about life itself before we can make a watertight case one way or the other. Many scientists who've thought about interstellar contact are excited about the prospect of learning from the accumulated knowledge of peoples who were launching space vehicles even before our Earth was born. But others have sounded a note of caution: in the unequal dialogue, we may come to emulate our superiors, and lose many of the qualities that make us human. For our pride's sake, it may indeed be better if we are alone in our Galaxy. Star captains of the future can proudly set sail for the farther reaches of the Milky Way, boldly setting foot on uncharted worlds, and claiming the entire Galaxy as the ultimate birth-right of mankind.

— 14 —

Beyond the Galaxy

What's the furthest you can see without a telescope? The answer – if you look skywards on a clear October night and pick out a misty oval patch in the constellation Andromeda – is an astonishing 21.5 million million million kilometres (13 million million million miles). That's the distance to the great Andromeda Galaxy, one of our Milky Way's *nearest* neighbours.

The nineteenth-century astronomers could see hundreds of these misty patches through their telescopes, but couldn't agree whether they were close – maybe even planetary systems in formation – or extremely remote. The matter wasn't finally settled until the 1920s, when astronomers using newer and larger telescopes managed to find Cepheid variable stars in some of the spiral nebulae. The Cepheids yielded distances almost vast beyond belief. Our Milky Way, hitherto regarded as the entire Universe, was rapidly downgraded to the status of 'just another galaxy'.

It wasn't even a very important one. With a span of 150,000 light years, the Andromeda Galaxy is half as large again, and it is estimated to contain some 400,000 million stars. Otherwise, the two galaxies are thought to look remarkably similar. Andromeda, too, is a slightly open spiral, but it's difficult to see this well because of its very tilted angle to us. However, astronomers in the Andromeda Galaxy may well be cursing the unco-operative angle of presentation of our own!

Because it lies 'only' 2¼ million light years from us, astronomers can pick out many kinds of objects in the Andromeda Galaxy that we have already met in our own. There are over twice as many globular clusters in its halo, and in its spiral arms we can see dark gas and dust clouds, glowing nebulae, young star clusters and even individual stars. The 'twin' illusion is helped along by the discovery that the Andromeda Galaxy, like our own, has two small close companion galaxies (both visible in a small telescope).

Most of the galaxies we see in the sky are spirals. However, they're by no means as common as it would appear from this: they stand out simply because they're all large and bright. All spirals contain thousands of millions of stars and have roughly similar sizes and brightnesses. Their variety comes in their endless range of shape and form, from tightly-coiled spring to wide-flung cartwheels. Photographs of spiral galaxies can fascinate for hours, and it's not hard to see how some of the most beautiful have developed names and personalities of their own which are used more widely than their 'proper' catalogue numbers. There's the glorious 'Whirlpool Galaxy' (M51), with its spiral arms arranged face-on in perfect symmetry, but tantalisingly distorted by a small companion. And the edge-on 'Sombrero Hat' (M104) is interrupted by such a dark swathe of dust that its upper regions in photographs look just like the crown and brim of a hat.

Spiral galaxies have smaller cousins, too: galaxies which have a similar mixture of young and old stars, dust and gas, but which seem to be too small to 'grow' spiral arms. They can come in practically any shape, and are called irregular galaxies for obvious reasons. The two closest – in fact, our Galaxy's two companions – are clearly visible in southern hemisphere skies as the Large and Small Magellanic Clouds. Both are quite a bit smaller than our Galaxy, but at a quarter the size of the Milky Way, the Large Cloud shows interesting signs of being a 'missing link'. It seems to have the beginnings of a spiral arm. And its central bulge, while not circular, is an elongated bar – rather like the subclass of spiral galaxies called barred spirals. One zealous astronomer has gone so far as to classify it as a 'one-armed Magellanic spiral'!

Like all irregular galaxies, the Magellanic Clouds show regions of extreme youth. Although astonomers believe that all – or almost all – galaxies are the same age, it seems that irregulars have only very slowly gotten round to the business of making stars. There are few old stars, and the spaces between the stars are clean and unpolluted by the dust grains they produce. The Magellanic Clouds are so transparent that you can see straight through them. But what they lack in dust, they make up for in gas: irregulars have cold hydrogen by the cloud-load, just waiting to give birth to stars. Not all of it is cold. The Large Cloud in particular is studded with glowing nebulae in which stars have just been born. In fact, the appearance of the whole galaxy is dominated by the tentacles of the vast Tarantula Nebula, whose convoluted traceries hide a star believed to be 2500 times heavier than our Sun and 100 million times brighter!

Although we don't know of any irregulars with more than a thousand million stars, there seems almost no limit to how small they can be. 'Dwarf irregulars' are only a few tenths the size of our Milky Way, with less than a hundred million stars, and we can only pick out those which live very close to us. But while they must be very numerous, the winner of the 'commonest galaxy in the Universe' title must go to their close relations, the dwarf ellipticals.

Quite distinct from spirals, ellipticals are the other main kind of galaxy in space. Whereas spirals (and irregulars) are young and vigorous, ellipticals are old and past it; their star-forming days were over long ago. With notable exceptions, most elliptical galaxies are featureless balls of old red stars with too little dust and gas ever to make stars again.

Dwarf ellipticals have the distinction of being the smallest galaxies in the Universe. They are hardly worthy of the title 'galaxy'; but with their million-or-so old red stars they make up quite self-contained little systems. Because of the similarity in the types and numbers of stars, some astronomers used to think that dwarf elliptical galaxies were no more than 'escaped' globular clusters. But these tiny galaxies seem to have a strong sense of their own identity. The few 'tramp' globular clusters closely resemble their 'bound' brothers – they may have expanded just a little in the freedom of open space. But the dwarf ellipticals are all of astonishingly low density. It's as if a cosmic giant had stamped on a globular cluster and smeared out its stars into the widest possible area. In fact, astronomers estimate that stars in these galaxies are a thousand times further apart from one another than stars in the neighbourhood of our Sun. So if there's anyone unfortunate enough to live in one of these almost non-existent systems, he would see only three stars in his sky on a clear night!

At the other end of the scale are elliptical galaxies which would easily qualify in a competition for 'the biggest galaxy in the universe'. Some of these giant ellipticals have over a million, million stars and distended haloes which stretch a million light years through space. More often than not, they're rogue galaxies, hotbeds of energetic and violent activity. We will be taking up their story in Chapter 15.

There seem to be as many kinds of galaxy as there are stars making them up. Can we learn more about them, as we did in the case of stars, by classifying and ordering them? The first astronomer to attempt this daunting task was the American Edwin Hubble. Hubble, incidentally, was a man who thrived on challenges. He turned to astronomy research after a Rhodes scholarship in Law at Oxford; he was a first-class sportsman (particularly in boxing, where he was good enough to be considered for the heavyweight championship); and during the First World War he achieved the rank of Major in the US Army. Not surprisingly, the Hubble classification of galaxies has stood the test of time since it was first introduced in the 1930s.

Hubble classed the galaxies simply by their appearance in photographs. He ordered the ellipticals along a sequence running from E0 (perfectly spherical galaxies) to E7 (very elongated galaxies) with the number representing the degree of elongation. He put spirals into two parallel sequences – those with round central bulges, the others with barred bulges – and divided these into three types each, a, b and c, depending on how tightly their spiral arms are wound. Our Galaxy, on Hubble's system, is classified Sb; some astronomers who think its arms are more open would put it nearer Sc.

Between the ellipticals and the spirals, Hubble found a kind of missing link: a disc-shaped galaxy without spiral arms which he labelled 'lenticular', or S0. And at the end of the spiral sequences, Hubble lumped all the irregular galaxies. The entire classification took on the look of a horizontal tuning-fork: 'Hubble's Tuning-Fork Diagram' it is to this very day.

Hubble's decision to classify galaxies just by their looks – and not for any deeper, more exotic reason – has really paid off. Almost all the galaxies

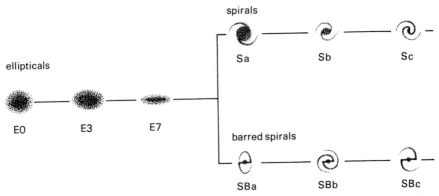

20 THE HUBBLE 'TUNING FORK' CLASSIFICATION OF GALAXIES *In the late 1920s, Edwin Hubble classified galaxies according to their appearance on photographic plates. This classification scheme has not only stood the test of time, but reveals real physical differences between galaxy types. Hubble's 'Tuning Fork' diagram places elliptical galaxies in a sequence of increasing elongation on the left, ranging from E0 (completely spherical galaxies) to E7 (very elongated galaxies): the number in each case compares the sizes of the major and minor axes of the galaxy. Hubble placed the spirals on the right, in two parallel sequences depending upon whether their nuclei were round ('normal' spirals) or bar-shaped ('barred' spirals). The suffix (a, b or c) refers to the shape of the arms: they denote respectively tightly-wound spirals with large nuclei, intermediate spirals (like our own Milky Way), and loosely-wound spirals with small nuclei. Hubble considered the possibility that his diagram might represent an evolutionary sequence, but it now seems that spiral and elliptical galaxies have quite separate characteristics right from birth.*

we know can be slotted in somewhere. But the Hubble diagram doesn't tell the same kind of story about galaxies as the Hertzsprung-Russell diagram tells about the stars. Hubble, understandably, thought it might. Perhaps galaxies started off as irregulars, picked up a bit of spin, sprouted spiral arms and gradually coiled them so tightly around themselves that they turned into ellipticals. Or maybe it happened the other way round.

Today, armed with better data than Hubble ever had, we know that galaxies can't spontaneously change from one type to another. Each type of galaxy has its own separate mass and size range. And *all* galaxies contain old stars, which tells us that all types must have been born at the same time. Galaxy formation is still such a mysterious and little-understood process that it gives away few of its secrets on Hubble's diagram. We shall see just how little astronomers know about galaxy birth in Chapters 15 and 16.

We do, however, know that galaxies like company. In the words of one Harvard College Observatory astronomer, 'Galaxies, like Legionnaires and Rhesus monkeys, are a gregarious lot.' Single galaxies, it seems, are rare; at the very least they are found in pairs, and the majority live in groups or even huge clusters of other galaxies.

Double galaxies, if close enough to one another, can look very spectacular. The Whirlpool Galaxy shows how two neighbouring star-systems can raise tides on each other, producing beautiful arcs and streamers of stars which curve away into the nothingness of space. Some double galaxies (usually called 'interacting galaxies') are actually colliding with each other. But neither suffers any ill-effects. Their stars are so far apart that the star-streams of each galaxy pass through one another like the ordered troops of a well-disciplined pair of armies. Only the gas clouds collide head on, leading to spectacular bursts of star formation as the clouds lose their energy and collapse. Sometimes this starbirth occurs well outside the confines of the parent galaxies, producing 'tails' of stars in intergalactic space. Pairs called 'The Mice' and 'The Antennae' are obvious examples!

Most of the galaxies we know live together in small groups. Our own Milky Way is no exception, belonging to a small collection of about forty galaxies which we call, predictably, the Local Group. It covers a flattened zone of space some five million light years across, and its three most important residents are the Andromeda Galaxy, M33 (a rather scruffy little spiral in the constellation Triangulum), and our own Galaxy. Its other inhabitants are mostly undistinguished dwarf galaxies which would hardly show up if they were any more distant – but it may just be able to lay claim to a giant elliptical, too. Unfortunately, Maffei 1 (discovered by Paolo Maffei in 1968) can be glimpsed only through one of the very dustiest regions of our Milky Way, and its distance is very difficult to determine. It's definitely a nearby elliptical galaxy, but the question is: how near?

Unspectacular though our Local Group may be, it does show us what an average region of space is like. It's something we need to bear in mind whenever we risk being carried away by the celestial fireworks of active galaxies and quasars! And the Local Group is not without its surprises. In recent years, radio astronomers have been discovering vast gas clouds close to some of its galaxies. One cloud completely surrounds both Magellanic Clouds and actually connects them to the Milky Way. In the opposite direction, it extends 200,000 light years into space as the 'Magellanic Stream'. Is it debris marking the path of the clouds as they approached our Galaxy? Or could it be the aftermath of a collision between the Milky Way and the Small Magellanic Cloud some 100 million years ago, in which the latter came off worst?

Other gas clouds appear to lie close to the Andromeda Galaxy and to M33. If they are at the distances of their associated galaxies – and unfortunately, there is no direct way of finding out how far away they are – then an interesting coincidence emerges. The clouds have roughly the same mass as the dwarf irregular galaxies. Could they, then, be the precursors of galaxies – galaxies which have yet to form? This idea is given some support by some recent discoveries of dwarf irregular galaxies (outside the Local Group) which really do appear to be making stars for the very first time. But a lot depends on how stable the clouds are. While there are astronomers who believe that the clouds will collapse to form stars, others maintain that their low density condemns them to expand forever.

We see a similar hotch-potch of spiral galaxies, dwarf galaxies and gas clouds in many of the other groups of galaxies nearby. None of these groups is very big, and they're all very typical of so-called 'irregular' clusters of galaxies.

'Regular' clusters, in contrast, are very different. While irregular clusters are small, shapeless and not terribly choosy about their galaxy types, the regulars show signs of immaculate breeding. They are all large. Some, like the Coma cluster, have thousands of members which swarm in a distinctly symmetrical way, reminiscent of a super-sized globular cluster, where each member is an entire galaxy, not just a star. This symmetry means that these clusters all look very similar. And this look-alike tendency is enhanced by their exclusive membership, which seems to place a ban on scruffy spirals – except perhaps around the outermost boundaries of the cluster. Otherwise, rich, regular clusters are almost entirely made up of elliptical and lenticular (So) galaxies.

Some of the ellipticals living in giant clusters have grown to a tremendous size. These supergiant ellipticals, never more than one or two per cluster, usually lurk at its very centre. On one classification scheme, they're labelled cD galaxies, which has led to the unfortunate nickname of 'seedy galaxies'! Seedy perhaps, but weedy, never: these supergiants contain over a million, million stars and their outer haloes stretch millions of light years into the cluster.

Astronomers reckon that the supergiants actually got that way by cannibalising other galaxies. Galaxies swarming near the tightly-packed centre of a rich cluster must suffer frequent traffic accidents. After a collision like this, a galaxy will lose all, or most of its energy, and it will fall to the very centre of the cluster where it doesn't have to move because it's at the cluster's centre of gravity. It will probably find a supergiant galaxy already lurking there, grown fat on the carcasses of similar unfortunate victims. To confirm this rather gruesome picture, astronomers point out that many supergiants appear on close examination to be made up of the remains of other galaxies, in particular that they contain material which is probably the cores of other galaxies.

These cosmic traffic accidents may be a source of the extensive gas found in rich clusters of galaxies. Unlike the cool gas clouds in the irregular clusters, this gas is extremely hot – at 100 million °C (180 million °F), it was actually first detected by the strong X-rays it emits – and it's concentrated on the cluster centre. Other astronomers believe it to be gas left over from galaxy-building. Either way it can provide an explanation for the absence of spirals in rich clusters, because a spiral galaxy would be scoured of its own gas as it moved through a relatively dense gas cloud. Spirals, then, can only live on the very outskirts of rich clusters where the gas density is low. Further in, we would expect to see lots of 'scoured spirals' – and that's just what astronomers believe lenticular galaxies to be.

The biggest clusters of all may even have more to them than galaxies and gas. In these giant swarms, galaxies are moving much more quickly than they should. It's almost as if the galaxies are having to rush around at breakneck pace in order to avoid being grabbed by the gravitational maw of some unseen matter. Astronomers have worked out from this that the biggest clusters should have ten times more mass than is visible – but where is the missing mass? Once again, we're back to the problems we encountered with the halo of our Milky Way Galaxy. Gas between the galaxies, even in these gas-rich giant clusters, can only account for a tenth of the missing mass. The other hypotheses include massive haloes around each individual galaxy, underluminous stars, black holes and the ubiquitous, possibly-massive neutrinos. Apart from slight evidence that the missing mass is spread throughout the cluster instead of being concentrated in particular places, astronomers aren't favouring any one idea above the others. Some scientists are hoping that the 'missing mass problem' will go away alto-gether as we continue to refine our measurements of galaxy speeds and luminosities. George Abell, whose sterling work in identifying over 2500 rich clusters when only a student led to its immortalisation as the Abell Catalogue, said recently: 'a re-examination of the data shows that a *serious* mass discrepancy does not *necessarily* exist in rich clusters.'

Single galaxies, double galaxies, groups of galaxies, clusters of galaxies – where does it end? Astronomers making detailed counts of galaxies all over the sky say that the buck stops at clusters of clusters of galaxies: superclusters. 'There's an average of 2–3 [giant] clusters per supercluster', says Abell. Mixed with the big swarms, we'd expect to find dozens of pairs and smaller groups of galaxies like the Local Group. A typical supercluster will cover perhaps 200 million light years of space, making these the biggest groupings of matter in the whole Universe. Our Local Group, and a number of other near neighbours, including the Sculptor Group and the Ursa Major Group, live on the edge of our Local Supercluster. This is a huge swarm of galaxies and groups of galaxies all centred on the giant Virgo Cluster of galaxies, some 60 million light years away from us.

We started this story of galaxies by thinking about distance, and it's with the same considerations that we end. Let us return to Edwin Hubble, whose pioneering work on galaxies, from his use of Cepheids as distance indica-tors, through his galaxy classification scheme, to the studies we are about to describe, 'undoubtedly rank', says one historian, 'as the most significant contributions to cosmology since the time of Copernicus'.

By the mid-1920s, Hubble had started on a program at the Mount Wilson Observatory to measure both the distances and velocities of the 'distant spiral nebulae'. If distance-determination was difficult, velocity measure-ments were even more so. Before the era of computers and image intensifiers, an astronomer had to spend gruelling hours photographing faint galaxies in order to get spectra which would show anything at all. More painstaking weeks in the laboratory would follow, as he attempted to measure the tiny Doppler shift in the wavelengths of the galaxy's spectral lines which would tell of its motion towards or away from us.

By 1929, Hubble had measured speeds and distances for eighteen galaxies; and he was puzzled. With the exception of the few very nearest galaxies, *all* the spectra showed a redshift: *all* the galaxies appeared to be moving away from us. The effect was systematic too. The more remote the galaxy, the greater was its redshift, which meant the higher was its recession velocity.

Over the next two years, Hubble attacked more galaxies, and increased his sample to over thirty. He was joined by an assistant, Milton Humason, formerly a donkey-driver on Mount Wilson who had become so fascinated by the astronomers' work that he stayed to help them. By 1934, they had measured over 100 velocities – and the same relationship between velocity and distance held. Hubble was able to predict that a galaxy twice as distant as one nearby will be moving away from us twice as fast; a galaxy three times further away will move three times as fast, and so on. In his honour, this relationship is now called 'Hubble's Law'. We can write it down simply as $V = Hr$, which means that the velocity (V) of a galaxy is equal to its distance (r) multiplied by a constant (H, now called the Hubble Constant). Today, it means that a measure of the redshift of a mystery galaxy will yield its distance.

When news of Hubble's work broke, theory and observation dovetailed together to provide an elegant explanation. Theorists such as Albert Einstein, Willem de Sitter, Aleksandr Friedmann and the Abbé Georges Lemaitre had been working on mathematical 'models' of the Universe's structure. The latter two scientists had independently predicted that the Universe should not be static but expanding, and that all the galaxies should be moving away from each other. And although Hubble was just a little cautious of this interpretation, it is this explanation which has stood the test of time.

Today, we can measure distances to galaxies thousands of times further away than Hubble could, and we still find that his law holds good. And because we have a much bigger sample to work with, we now know its more accurate to say that it's the *clusters* of galaxies which are moving away from each other, rather than the individual members of clusters. This means that clusters of galaxies are permanent, but that superclusters will one day cease to exist as their members move apart.

This explanation of the redshift, as being due to the expansion of the Universe, is accepted by the vast majority of scientists – but it is still open to debate. In the last few chapters, we shall meet objects which openly seem to defy Hubble's Law, and which some astronomers use as ammunition to back up their claim that the redshift needs to be looked at anew. We shall

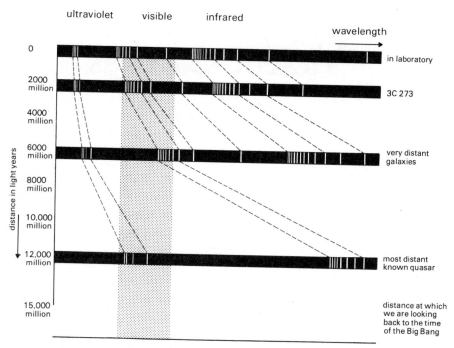

ultraviolet   visible   infrared

wavelength

21 SPECTRA SHOWING THE REDSHIFT OF THE GALAXIES *In the same way that the Doppler effect makes a rapidly-receding ambulance siren drop in pitch, so the expansion of the Universe stretches the light waves from distant objects towards the long-wavelength part of the spectrum. The light from almost every object beyond our Galaxy is said to be 'redshifted', and the amount of the redshift is very accurately proportional to the object's distance from us. The spectral lines (of hydrogen in the diagram) are stretched in proportion to their original wavelengths, and so longer-wavelength lines are shifted more. However, the ratio of wavelength increase to the original wavelength is the same for all lines in the spectrum of any particular object. Redshifts offer a way of measuring distances to the most remote objects in the Universe, which lie well outside the range of conventional techniques.*

also find that the tools of those scientists who investigate the Universe's largest scales and deepest problems aren't simply huge telescopes, computers and particle accelerators, but include an appreciation of order, a sense of wonder, and a feeling of humility for the awesomely simple power of Nature. The last story sums it all up.

Just after Hubble's discovery of the redshift-distance relationship, Albert Einstein and his wife Elsa came to visit him at Mount Wilson, where they were given a tour of the Observatory. An astronomer showed them the giant 100-inch Hooker telescope (described as a 2.5-metre these days) and explained that it was used for determining the structure of the Universe. It wasn't Albert who replied, but Elsa. 'Well, well!' she exclaimed. 'My husband does that on the back of an old envelope.'

— 15 —

Violent Galaxies

The vast majority of galaxies are orderly, law-abiding citizens of the Universe. But every community has its offenders, and the Universe, too, has some inhabitants which are disturbed, delinquent or just downright violent. Their numbers are very small. But because these 'active' galaxies (as they're generally called) are amongst the brightest and most powerful in the Universe, they positively shout out to be investigated.

Astronomers hardly had time to get used to the idea of 'normal' galaxies before the first of the 'active' variety reared its dazzling head. A young astronomer called Carl Seyfert, working in the USA in 1943, drew attention to a small number of spiral galaxies – about a dozen – which seemed to be behaving in a pretty strange way. At first he had thought that their spiral arms were very faint, but he then realised that this was simply a contrast effect – the centres of these galaxies were in fact unusually bright.

As well as being so bright, their centres looked very odd, appearing as tiny, starlike points instead of the well-developed central bulges of most spiral galaxies. Intrigued, Seyfert took spectra of his galaxies. He found the cores of these galaxies did not show the spectrum of a mass of stars, but were dominated by the spectral lines of hot gas, rushing around the centre with considerable speed.

Because they were such an anomaly at the time, no one else joined Seyfert in his researches into these strange galaxies. His findings remained practically unknown until the early 1960s, by which time astronomers, having come across many more inexplicable galaxies, had resorted to ploughing through past literature. Seyfert himself had died just beforehand, but astronomers honoured his memory in giving his name to this whole new class of galaxies.

The several hundred Seyfert galaxies now known are basically giant spirals like our own, but they have cores (or nuclei) which are extremely disturbed. Hot gas at 20,000 °C (36,000 °F) sweeps about the brilliant nucleus with speeds of up to 7000 kilometres per second (that's 16 million miles per hour!). Sometimes, the disruption has spread to the outer parts of the galaxy. The Seyfert galaxy NGC 1275, with its outward-moving filaments of gas, looks like a gigantic version of the Crab Nebula (see page 107).

The core itself gives out a tremendous amount of energy over the whole range of the wavelength spectrum. Its light output dominates the rest of the galaxy, and the same is true for the X-rays and infrared radiation it produces. A few Seyferts give out radio waves too. All of this points to the core being a site of violent and extreme conditions; it is the 'powerhouse' which gives energy to 'a dragon in the middle'.

Astronomers would like to know what the dragon *is*. They know that it is small, for the entire disturbed region surrounding it is only a few tens or hundreds of light years across. But the very degree of its smallness came as a surprise, bringing with it the first of many energy problems that were to confound researchers investigating active galaxies.

The initial difficulty is that the core is too small for any telescope to photograph. Just as in the case of a bright star, the continual shifting of our atmosphere smears out its point-like image into a much bigger disc which has nothing to do with its *actual* size. But astronomers 'hunting the dragon' got help in their search from a totally unexpected quarter: old star catalogues. Photographic catalogues made around the turn of the century had erroneously classed some of the brighter Seyfert galaxies as stars. All that the earlier catalogues could see was the brilliant, starlike core without the faint extensions of the surrounding galaxy, and their mistake was perfectly understandable. What was particularly interesting was that many of these 'stars' had been classified as variable, changing irregularly in brightness by a factor of two over a period of a few months to a year.

Measure the time for an object to change in brightness, and you can get an estimate of its size. Imagine a glowing ball of gas, measuring a light year across. Now think of it being abruptly 'switched off'. The ball itself won't go dark immediately, however, because light itself takes time to fade away across it. The surface closest to you will go dark first, but it won't be until a year later that the back surface – a light year further away – will fade. In other words, a light source can't change in brightness any more quickly than the time it takes light to cross it.

If the nuclei of Seyfert galaxies can vary in brightness on timescales of a few months, it means that they must be only light-months across. This really is tiny on the scale of a galaxy, and more comparable with the scale of our solar system (Pluto's orbit spans eleven light-hours). Some nuclei may be even smaller than this. There's one whose X-ray output, as monitored by the orbiting Einstein Observatory satellite, varies on a timescale of $2\frac{1}{2}$ minutes. This means that a significant proportion of its energy is coming from a region less than half the size of Mercury's orbit about the Sun.

The size of the core would be no problem if it weren't for the energy it generates. Seyfert galaxies lie nearby, and it's relatively easy to measure their distances and convert the apparent brightness of their cores into real luminosities. This luminosity corresponds – very nearly – to the total energy of a typical galaxy like ours packed into a region a few sizes larger than our Solar System.

Faced with an energy problem like this, the reaction of many early-1960s astronomers was to hope it would go away and lose itself. Only, it didn't. It actually got worse, as subsequent discoveries highlighted galaxies with even greater reserves of power.

Before we face these galaxies – the most violent in the Universe – there are a few ends to be tied up. The Seyfert galaxies we see are highly energetic at the moment, but they couldn't possibly have kept up this profligate output for any more than a small fraction of their lives. Otherwise, there would be no galaxy left to see! This means that Seyfert activity must be intermittent. The question is: how frequently does it happen?

Earlier on, we mentioned that Seyfert galaxies are, by and large, giant spirals. In fact, one in ten of all giant spirals is a Seyfert. We can turn this statistic around to ask: is it possible that *all* giant spirals spend one part in ten of their lives undergoing Seyfert outbursts? In particular, has *our* giant spiral spent ten per cent of its lifetime like this? As we learnt in Chapter 11, the available evidence suggests that it has.

Seyfert activity almost pales into insignificance as we ascend the ladder of extragalactic violence. On the next rung up, we find the galaxies whose identification, in the 1950s, helped to unearth Seyfert's forgotten spirals. These radio galaxies were different again. They give out as much energy in radio waves as they do in light, and some of them are tremendously powerful. The radio galaxy Cygnus A is actually the second brightest radio source in the whole sky – yet it lies well outside our Galaxy!

Radio and optical astronomers collaborated together in an attempt to pin down the kind of object which could emit radio waves thousands of times more strongly than any normal galaxy. Co-operation between the two camps was a little strained to begin with. For a start, it was found that a few radio astronomers, trained as pure physicists, were plotting their sky co-ordinates back-to-front in what appeared to be a more logical way! But co-operation eventually paid off. A handful of most peculiar-looking galaxies were isolated as sources of the radio waves, and the optical astronomers began their crackdown to discover what kinds of galaxies these were.

Cygnus A was typical. It was clearly very distant, but showed a distinctly double shape, like a pair of galaxies colliding with each other. Perhaps *this* was the origin of their powerful radio waves? Then maybe there would be other signs of such a titanic disturbance, such as streams of hot gas. Leading observer Walter Baade – the man who doubled the size of the Universe at a stroke (Chapter 11) – had a hunch that a spectrum would show these up. His long-time colleague Rudolph Minkowski was equally adamant that Cygnus A was not a pair of colliding galaxies, and predicted that hot gas wouldn't be detected. The men made a famous bet together with a bottle of whisky as the stake. Baade spent all night on the 200-inch Palomar telescope (now called the Hale 5-metre) in an attempt to secure the spectrum of the dim and distant galaxy, or galaxies. He got his bottle. There on the spectrum were the tell-tale emission lines from hot, excited gas: proof at last that radio galaxies were galaxies in collision.

Baade's hunch was a good one; but unfortunately, it was wrong. Other astronomers were not slow to realise that a pair of colliding galaxies could never unleash as much energy as radio galaxies do. As we now know (Chapter 14), galaxy collisions are fairly peaceful affairs.

Our present ideas about radio galaxies are likely to be much nearer the mark. It's not because scientists are any more brilliant, but a result of changes in the pattern of science over the past few decades. Ironically, increasing specialisation in science has led to more co-operation between scientists of all disciplines, perhaps in an unconscious attempt to preserve the overall picture. International science today means sharing facilities, whether computers, telescopes or simply ideas bounced between participants at conferences and 'workshops'. Astronomers don't tend to put labels on themselves any more. Instead of being 'optical astronomers' or 'radio astronomers', they're all physicists who use the Universe as their laboratory.

Nowhere has the spirit of co-operation been more marked than between the observers and the theorists. It's fair to say that much of today's revolution in astronomy has been brought about by observers and theorists working hand-in-glove, each group spurring the other on. It wasn't always so. Observers have traditionally been suspicious of scientists who seemed to deal in esoteric concepts and with scant regard for reality. But time and time again, whenever observers found something new and inexplicable they also found that the theorists had been quietly, and independently, working away on the same problem. After the round of convincing explanations for the expanding Universe, the microwave background (Chapter 16) and pulsars, observers and theorists realised that their interests were best served in joining forces. As leading theorist Tommy Gold said in an after-dinner conference speech in 1963, welcoming relativity specialists to the astronomers' camp: 'It was ... [Fred] Hoyle's genius ... that allowed one to suggest that the relativists with their sophisticated work were not only magnificent cultural ornaments but might actually be useful to science!'

Theorists and observers alike agree that radio galaxies are a still-more powerful version of an exploding galaxy. The galaxies responsible for the immense radio output often look disturbed, sometimes slashed with dramatic swathes of black dust generated by their outbursts. Many of them contain streams of gas moving with speeds in excess of 6000 kilometres (3700 miles) per second. But for most, the hallmark of their violence is located in the regions that generate their radio waves. Almost all galaxies emit some radio waves, but these come from the galaxy itself, from the gas which lives between the stars. Radio galaxies are quite different. Although some radio emission arises in the main body of the galaxy, most of it comes from two gigantic lobes of gas which straddle the galaxy in a roughly symmetrical way. These clouds have been somehow 'beamed out' by the nucleus of the galaxy, a sure sign of explosive activity. A large number of radio galaxies have jets of matter – sometimes visible optically – spurting up to three million light years from their nuclei into the radio-emitting lobes. 'Such jets have become common, almost ubiquitous' say Martin Elvis and Andrew Wilson, writing in the scientific journal *Nature* on discoveries reported at a recent conference on radio galaxies.

Like Seyferts, the radio galaxies pose considerable problems when it comes to their energy supply. Somehow, they are able to spew out matter from their cores at a tenth the speed of light – and not just in large quantities, but out to considerable distances, too. Many radio galaxies have clouds which span more than a million light years. The present record holder is a galaxy called 3C 236 (the 236th source in the Third Cambridge Catalogue of radio sources), whose clouds cover an astonishing 18 million light years – some nine times the distance between the Milky Way and the Andromeda Galaxy. Even so, it mightn't be the biggest of all. Some radio astronomers believe they have detected extended clouds around a few otherwise moderately large radio galaxies – and 3C 345, the biggest of all, could be 78 million light years across!

The clouds themselves can give some clues about the powerhouse at the centres of radio galaxies. By measuring the amount of energy they give out at different wavelengths, radio astronomers can plot an 'energy spectrum' to find what makes the clouds glow. They have discovered that the radio waves are emitted by tiny, negatively-charged electrons, which have somehow been accelerated to velocities close to the speed of light. When these fast-moving particles meet the weak magnetic fields of intergalactic space, they are forced to spiral around the field lines and give out their energy in the form of synchroton radiation – so-named because it was first detected in powerful synchrotron particle accelerators on the Earth. The energy is emitted at radio wavelengths, but in some radio galaxies it is so intense that the clouds can be faintly seen glowing at optical wavelengths too.

The clouds' shapes reveal still more about these strange galaxies and their surroundings. Quite often, their structure is very complicated, with powerfully-emitting 'hot spots' marking where the beam is still feeding energy into the cloud. Some galaxies' twin clouds have been practically swept back and merged into a trailing tail as the galaxy has moved through the relatively dense gas in a cluster of galaxies. These look irresistibly like squirming tadpoles, and the name 'tadpole galaxy' has stuck.

Most important of all, many radio galaxies are surrounded by more than one cloud system. This is firm evidence for repeated activity, with the larger clouds produced by an earlier burst of energy than those closer to the visible galaxy. These nested-cloud galaxies tell us that the bursts of activity which produce the clouds last for a few million years. After such an outburst, the parent galaxy becomes dormant for 100 million years or so. Then something triggers the central powerhouse to repeat the act all over again. It is astonishing that, depite the 100 million years dormancy, the repeat outburst occurs in almost exactly the same direction as its predecessor. When it comes to tracking down the nature of the powerhouse, it's this sort of evidence which gives the theorists something to bite on.

What kind of galaxy suffers convulsions such as these? The only sort which can generate activity on this scale are giant elliptical galaxies with the power to muster a huge concentration of mass towards their centres. Many of the cD galaxies lurking at the heart of rich clusters are powerful radio galaxies, possibly as a result of the gas fed to them by their cannibal life-style. Less easy to understand are strong radio galaxies, like the devastated Centaurus A, that live completely alone. What's certain, however, is that there's no chance of our Milky Way turning into a radio galaxy. Spirals are too small ever to develop the extreme central densities that a radio galaxy's powerhouse evidently needs.

Even the violent radio galaxies are dwarfed by the quasars. These strange objects – insignificant to look at, and yet incredibly powerful – are still giving astronomers headaches twenty years after they were first identified. The quasar story began in the same way as the hunt for the radio galaxies, when radio astronomers located a number of powerful sources of radiation in the sky. But this time, try as they might, the optical astronomers just couldn't find the culprits responsible. Every time, the relatively imprecise radio position was occupied by a field of thousands of apparently ordinary stars. There was no dust, no gas and nothing noticeably peculiar.

The astronomers, in desperation, turned their attention to the myriads of faint stars covering the radio position. They carefully checked out every one for signs of abnormality and at last their patience paid off. In each field were one or two stars which were definitely too blue to be 'normal'.

The next stage was to get spectra of these stars, to check on their temperature and composition. There is no reason *why* a star should give out powerful radio waves, and so the puzzled astronomers were expecting a very strange kind of spectrum indeed from these mystery objects. But they were totally unprepared for what they got: in every case, a spectrum of an evidently very hot object, crossed by spectral lines that couldn't be pinned down to any known chemical element.

The astronomers of the early 1960s were totally baffled. The objects were discussed at the December 1960 meeting of the American Astronomical Society, and the conclusion astronomers reached about 3C 48 – one of the mystery stars – was that 'there is a remote possibility that it may be a distant galaxy of stars; but there is general agreement among the astronomers concerned that it is a relatively nearby star with most peculiar properties.'

Nature co-operated to confirm that at least one 'quasi-stellar object' (or 'crazy stellar object', in the words of the young daughter of astronomers

Margaret and Geoffrey Burbidge) was definitely the source of strong radio waves. In 1962, the Moon conveniently passed in front of one of the objects, 3C 273, and simultaneously both light and radio waves were cut off. Astronomers could at least relax in the knowledge that their blue stars were leading them along the right trail.

The last instalment of the discovery story involves not so much astronomy, as scientific instinct – the 'aha!', as astronomer Charles Long describes it. When Maarten Schmidt of the Hale Observatories sat down in 1963 to write a report on 3C 273, all the jumbled pieces of the puzzle were there, but no one had been able to fit them together. 3C 273 looked like a star with a jet coming out of it; it was the brightest of all the quasi-stellar objects (although only magnitude 13), and its spectrum was reasonably clear. Schmidt's subconscious mind must have turned to that peculiar spectrum again and again. There was something familiar about it: to an expert spectroscopist like Schmidt, an oddly recognisable pattern of lines.

Then he realised. The lines in the spectrum were a part of a well-known series of lines produced by hydrogen atoms, but all shifted to longer (redder) wavelengths. The longest wavelength lines of the series were missing – shifted out of the visible altogether, and into the infrared part of the spectrum by the object's enormous redshift. Schmidt measured the redshift and found it to be 0.158, corresponding to a recession velocity of 15.8 per cent of the speed of light and – from Hubble's redshift-distance law – a distance of almost 2000 million light years. The 'star' was as distant as the most remote galaxies known. And judging from its apparent brightness, it was much more luminous.

The other quasi-stellar objects (a term mercifully shortened to 'quasars') turned out to have higher redshifts still. That of 3C 48, the subject of discussion at the 1960 AAS meeting, came to 0.3675, placing it 4500 million light years away with a recession velocity of 36.75 per cent of the speed of light. The highest quasar redshift yet measured belongs to PKS 2000–330. This quasar was discovered by Australian and British astronomers in 1982: its redshift of 3.78 means that it is rushing away at over 90 per cent of the speed of light, and lies at a breathtaking distance of 13,000 million light years.

Today, over 1500 quasars have been found. The majority, as it happens, do not emit radio waves, making their original discovery and identification something of a lucky accident. With a few exceptions, they look exactly like faint stars. It is only their enormous redshifts – and in some cases, strong infrared and X-ray emission – which give them away.

Beneath a quasar's undistinguished façade lurks one of the Universe's most potent power generators. Energy is packed into the quasars to a truly astonishing degree. Like Seyfert galaxies, many quasars vary in their light output, which, as we have learnt, tells us something about the size of the source. But these celestial beacons can change in brightness by a factor of ten, in just a matter of months, while their close relatives, the BL Lac objects, sometimes flare up to become a hundred times brighter over the same kind of period. X-ray astronomers report that some quasars flicker on timescales of seconds. This all points to quasars being extremely small. They cannot be much larger than our Solar System, but within that minuscule region they pack the total energy of no less than a hundred galaxies. Energy is concentrated to an extent a million, million, million times greater than it is in any part of our Milky Way Galaxy. It scarcely sounds believable. And this was exactly the reaction of many 1960s' astronomers to this outsize energy problem. Even today, there are some who still strive desperately to make it go away.

One way of killing the energy problem at a stroke is to postulate that the quasars are nearer to us. Their size won't change – their short variability timescale still puts them in the solar system size class – but their energies will. If they are nearby, they don't have to be nearly as luminous. But surely their enormous redshifts automatically consign them to vast distances? This, argued some astronomers, supposes that the redshifts of quasars are due to the expansion of the Universe – but quasars are such peculiar objects that it's possible that even their redshift has some other origin. Isn't it just as likely, they suggested, that quasars are nearby objects with high redshifts produced by physical processes *inside* them?

One idea could be scotched very quickly. It was possible that quasars were fast-moving objects ejected from galaxies, and their redshifts arose from their high speeds – but in this case we would expect to find some blueshifts too, and quasars never show a blueshift. The only other serious alternative is that the redshift of quasars is due to their gravity. Light, as a form of energy, is affected by gravitational fields in the same way that material bodies are. To escape from any gravitational field, a body has to use up energy: the stronger the field, the more energy has to be expended. In the case of light, the energy used shows up as a redshift, and the stronger the gravity, the greater the redshift. Light from dense, high-gravity stars like white dwarfs shows a marked gravitational redshift. So perhaps quasars are dense objects which lie nearby and there is no excess energy problem to explain? Or does an explanation of their redshift require some 'new physics', as a few astronomers have ventured to suggest?

The 'local quasar' astronomers, although more vociferous in the 1960s, have scored some apparent victories in the past few years. They point to some quasars apparently spewing out blobs of matter with several times the velocity of light – something which contravenes the theory of relativity. This problem is removed if quasars lie close at hand. Ace quasar-hound Halton ('Chip') Arp has highlighted several alignments of quasars in the sky which he claims add weight to his 'local ejection' hypothesis. In addition, Arp has spent the last two decades looking for physical connections between obviously nearby galaxies (of low redshift) and quasars (of high redshift). He reports a greater-than-expected incidence of quasars near galaxies and close quasar galaxy pairs – and in each case, the galaxy and quasar have wildly different redshifts. In a few cases, a faint bridge of stars linking the two seems dimly discernible.

It's fair to say that most astronomers feel that Arp is barking up the wrong tree. As we shall see later, there's really no need to make quasars local, or to search for some exotic explanation of their redshifts, for the 'energy problem' can be explained in an elegant and consistent way. But at first sight, the 'local' evidence is persuasive. However, other astronomers have pointed out that the apparent 'faster than light' velocities in quasars – so-called 'superluminal expansion' – can be easily explained by fairly simple geometry. If a blob or jet from a quasar is coming at us almost head-on at a speed close to that of light, detailed calculations show that it will actually seem to move faster than light. As the *Los Angeles Times* somewhat clumsily reported: 'US Scientists Measure Super Velocity of Quasar and Conjecture It Is Illusion'.

Arp's apparent alignments look set to follow the same road. Cardiff astronomer Mike Edmunds has used the University's computer to simulate a random sprinkling of quasars, and finds that fifteen lines of three quasars each will occur by chance in any given region of the sky.

But what of the most persuasive evidence of all – the apparent connections between nearby galaxies and allegedly remote quasars? Once again, the 'traditionalists' have an answer. Some of Arp's 'bridges' have been re-examined using today's improved scanning and computing techniques, and it seems that the majority of them really don't exist. Most are simply contrast effects between two densely-grained and narrowly separated areas of photographic emulsion.

However, it's not that easy to dismiss the apparent clustering of quasars around galaxies. Although it's a very difficult statistical problem to predict how many quasars should be found in any given area of sky, there really does seem to be a marked increase in their numbers close to galaxies – just as if they had been thrown out of the galaxies themselves. But even this puzzling state of affairs can be explained in a conventional, if seemingly bizarre, way.

As we already know, gravitational fields rob escaping light of its energy to produce a gravitational redshift. Gravitational fields can also bend light like a lens, causing almost invisible objects to be intensified into bright, often multiple images. This strange effect is demonstrated *par excellence* in the 'double quasar': two quasars of exactly the same redshift separated in the sky by a distance equivalent to only one three-hundredth the diameter of the Full Moon. The similarity between the two quasars is too good to be true, and astronomers quickly concluded that they were observing the *same* object, split into two by some invisible gravitational lens. Hunch is all very well, but proof didn't come until astronomers using a sensitive electronic light-detector on the Hale 5-metre telescope were able to photograph an extremely faint galaxy lying between the two quasar images. This galaxy lies very much closer to us than the quasar, and its gravitational field focuses its light into two (and probably more) surrounding images.

This peculiar lens effect can explain the concentration of quasars around galaxies, particularly so if the latter have massive haloes. According to George Canizares of the Massachusetts Institute of Technology, 'the focusing can intensify and make detectable quasars which would otherwise have been below the threshold'. In this case, the quasar images aren't split by the gravitational lens, but are simply intensified by it. In regions away from galaxies, the quasars remain as faint as ever.

And so it seems that the quasars really do lie at the distances indicated by their huge redshifts. They really *are* phenomenally distant and uncompromisingly energetic. 'Why fight the evidence?' is the attitude of many scientists. 'Many times in their 20-year history, quasars have been charged with breaking the known laws of physics in one way or another, and each time they have been cleared', says *New Scientist* magazine in their defence.

But just what *are* quasars, anyway? It's a question that has challenged astronomers for many years. Only in the last decade has a consensus begun to emerge, following a concerted attack from many of the world's topflight scientists who were goaded into action by the quasars' apparent disregard for conventional physics. Astronomy is far from being a field where people simply 'make discoveries'. One of its main concerns is to test whether the physical laws which we know to hold good on Earth also apply in the far more extreme conditions of space. An astronomer uses the Universe as a laboratory more versatile than any he could find here: where he can test the behaviour of matter under conditions of excessive heat and cold, high and low density, ultra-high velocities and on timescales even longer than the age of the Earth. The most tempting and challenging testbeds are those objects, like quasars, which are the most extreme of all.

Alec Boksenberg, Director of the Royal Greenwich Observatory in Sussex, is one of the world's leading quasar investigators. Fascinated by tantalisingly remote, but dismayingly faint galaxies and quasars, he has developed an ultimate light-gathering device – the Image Photon Counting System (IPCS) – to wring their feeble light-waves dry of information. Quasars and galaxies yield up their secrets in their spectra, which tell astronomers of their composition and physical conditions. But quasars are so faint that early quasar spectra, recorded on photographic plates, were sometimes prohibitively difficult to decode with confidence. Now astronomers using Boksenberg's IPCS and other, similar electronic gathering systems – far more efficient at recording elusive light-waves than a photographic plate – can watch a minicomputer at the end of a telescope build up a quasar's spectrum in unprecedented detail in 'real time' (a technical phrase meaning that input and output are simultaneous). The astronomer can interact with the device if he wants to concentrate on any particular feature – perhaps a faint spectral line of special interest. But the real advantage of electronic light gathering detectors lies in their exceptional speed. While in 1929 it took Milton Humason 45 hours on the 100-inch (2.5-metre) Mount Wilson telescope to get a spectrum of a faint galaxy – a task which he had to spread out over five consecutive nights – a modern image tube spectrograph can do the same job in ten minutes!

Sophisticated electronic cameras and subsequent computer image processing have also helped to rob quasars of some of their mystery. A number of the nearer quasars (on the basis of their redshifts) have now been found to be surrounded by a very faint 'fuzz' which appears to be a surrounding galaxy almost totally drowned out by light from the dazzling 'quasar' core. There's a growing body of evidence which suggests that quasars are vastly scaled-up versions of Seyfert galaxies. Husband and wife team Sue Wyckoff and Peter Wehinger maintain that 'If . . . quasars are high-redshift analogues of Seyfert galaxies, then we should not be too surprised that they appear stellar even on long-exposure photographs.' Only in the very nearest quasars can we pick out the dim galaxy which surrounds the tiny, disturbed nucleus.

Quasars, it now seems certain, are yet another kind of violent galaxy. While their spectra look like those of Seyferts – interrupted by spectral lines arising in hot, fast-moving gas streams – they behave, in some cases, more like a radio galaxy. But they're more powerful than both, and the most luminous objects by far in the Universe.

In all this wanton violence amongst galaxies, there is an underlying pattern. Violence occurs only in galaxy cores, where the density, temperature and pressure are sufficiently high. And violence is closely associated with distance away from us. Nearby galaxies, in the main, are undisturbed – with the exception of the slightly active Seyferts. In the realms of the medium-to-far distant galaxies, we find the badly-disturbed radio galaxies, while awesomely-powerful quasars are found only in the most remote reaches of the Universe.

All this may sound quite arbitrary until we think of it this way: as well as being segregated in space, active galaxies are segregated in time. Because Seyfert galaxies lie nearby, their light takes a relatively short time to reach us and so we see them very much as they are in the present. The light from radio galaxies can take several thousands of millions of years to reach us, and so we see them not as they are today, but as they were thousands of millions of years ago – in other words, when they were quite a lot younger. In quasars, the effect of this 'lookback time' is the most dramatic of all. A quasar 10,000 million light years away will look to us as it did 10,000 million years in the past, a mere 5000 million years after the beginning of the Universe. Quasars, then, are among the earliest objects we can see – and the most violent.

Piecing together the evidence, astronomers have come to the conclusion that it's not just human beings who are prone to periods of violent behaviour during youth. Galaxies appear to be similarly affected. Far from being a random or unexpected phenomenon, it may be that all galaxies go through an active phase when they are young, and that galaxies which are currently active can tell us something of the way in which most galaxies grow up.

No one has yet seen a galaxy being born. But as we shall see in Chapter 16, there's good evidence that they begin their lives as vast gas clouds which collapse ever more swiftly under the relentless pull of gravity. And at the centre, contraction occurs fastest: so quickly and dramatically that the young galaxy's core collapses in on itself to become nature's surest gravitational trap – a massive black hole.

A 'galaxy-mass' black hole is on a totally different scale from those created in stellar explosions (Chapter 10). Depending on how much matter there is available in the 'protogalactic' gas cloud, a young galaxy may be able to 'grow' a central black hole up to thousands of millions of times the mass of an average star. It continues to grow as fodder from the still-collapsing galaxy pours in. But young black holes are messy eaters, and not all the material is gobbled up at once. The residue collects in a vortex-like 'accretion disc' surrounding the hole, glaring and flickering fitfully as its globs of gas sweep around at speeds close to that of light before they're swallowed up forever.

The accretion disc glows fiercely, giving out alarming quantities of energetic, short-wavelength radiation. It's probably this flaring beacon that we call the core of a Seyfert galaxy, or a quasar. But the accretion disc does more than just glare. It is the 'dragon' at the centre of the galaxy; one of the most potent generators of power known in nature.

On the present scenario quasars are the accretion discs at the centres of young massive galaxies. These swirling discs generate powerful shockwaves which send gas clouds hurtling through the young galaxy, but they can act in a more organised way too. Brilliant Cambridge theorist Martin Rees has demonstrated that a massive black hole and accretion disc combination can shoot beams of highly energetic particles along its axis of rotation until they end up as huge clouds which straddle the galaxy in space.

Although all quasars appear to have a 'dragon', only a few are surrounded by clouds. It may be that this is a late stage in a quasar's development. But it seems to link it to the next phase in the life of a massive elliptical galaxy, that of a radio galaxy.

Are radio galaxies actually dead quasars? As we've seen, the links seem strong, and radio galaxies certainly have disturbed or slightly active cores. Many have beams emanating from their cores which feed the surrounding clouds. And the orbits of stars near the centre of at least one radio galaxy seem to be governed by the presence of some unseen mass more than 5000

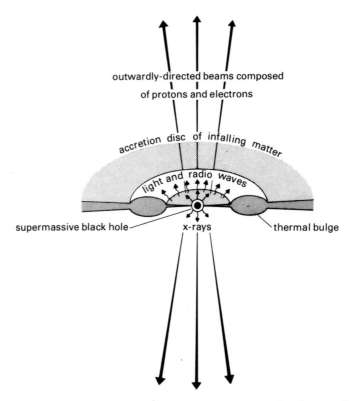

22 THE POWERHOUSE OF A RADIO GALAXY OR QUASAR *Is this the powerhouse of a radio galaxy or quasar? Many astronomers believe that the only way to explain the colossally-powerful and incredibly tiny cores of these objects is to draw on the enormous gravitational energy of a supermassive black hole. Matter spiralling into such a black hole – thousands of millions of times the mass of our Sun – would form a glaring, doughnut-shaped accretion disc which pours out radiation at all wavelengths. Physicists have calculated that a whirling accretion disc like this can shoot off beams of high-speed particles along its rotation axis. These beams later billow out into voluminous radio-emitting clouds, millions of light years apart and straddling the galaxy in space.*

million times the mass of our Sun. But it's the repeated activity in radio galaxies which points to a black hole at work. As we saw earlier 'explosions' in radio galaxies are separated by hundreds of millions of years, and yet even after this length of time, the core appears to 'remember' the direction of its previous outbursts. At the moment the only contender for the title of 'cosmic elephant' is an accretion disc. When sufficient material falls on to it, the disc is triggered into beaming along its axis of rotation – which always points in roughly the same direction.

Radio galaxies, too, must die. Leading optical astronomer Jim Gunn points out that there are enough problems involved in 'feeding the monster' at the centres of these notoriously gas-free galaxies anyway – and the day must come when the accretion disc can be fuelled no more. It's on the cards that some nearby giant ellipticals could well be dead radio galaxies, which, in their profligate youth, were quasars.

Seyfert galaxies are the poor country cousins of quasars and radio galaxies. As less-massive spiral galaxies, they were never able to develop the truly heavyweight black holes which may lurk at the hearts of their elliptical counterparts. As a result, their acts of violence have always been more limited; although their cores can generate shockwaves, they were never powerful enough to beam effectively. But because Seyferts are gas-rich spirals, the 'dragon' still has food, and activity, albeit mild, still continues today. Even, perhaps, in our own galaxy.

The last word goes to dragon-inventor Daniel Weedman: 'We have … an incredible phenomenon that can range over a factor of a million in luminosity without changing its properties … the inference … is that the quasar phenomenon is an event in a galactic nucleus, low-luminosity examples of which are given by the Seyferts.'

At first sight, a fascination with galactic violence might seem a rather macabre preoccupation for the best brains in the astronomical community. But as we have seen, it's one of the most challenging areas of astrophysics, and a decisive test-bed for physical laws. It tells of the birth and evolution of galaxies, of matter and radiation in the most extreme conditions, and of the harrowing violence of our early Universe. As we, the products of a calm, mature Universe contemplate the serenity of our surroundings, those tranquil galaxies upon which we rest our gaze may well be relishing fond memories of a turbulent youth.

— 16 —
The Big Bang

'We frankly did not know what to do with our result', despaired Arno Penzias, '... "knowing", at the time, that no astronomical explanation was possible.' Hardly a promising start to a finding which ranks with Hubble's discovery of the expanding Universe as the most important in astronomy this century. It went on to win Penzias and his colleague Robert Wilson a share of the 1978 Nobel Prize for Physics – and world fame as the two men who had shown how our Universe began.

Any thoughts of fame or discovery must have been far from Penzias's and Wilson's minds as they started their research project in the early 1960s. They weren't even studying the origin of the Universe. Instead, they were keen to map the radio waves coming from directions away from the main plane of our Galaxy – measurements which needed a particularly sensitive radio telescope to pick up the faint background 'noise'. Any electrical disturbance in the telescope's circuits would drown the weak signal, and so the two astronomers had to go in search of an exceptionally low-noise receiver. Their hunt brought them to an unusual 6-metre (20-foot) horn reflector with ultra-low noise. This had been specially built for work with the old *Echo* communications satellite, and it was based at the Holmdel, New Jersey, site of the Bell Telephone Laboratories. Quite coincidentally, this was precisely the place where Carl Jansky accidentally discovered radio waves from space in 1931.

To test out the receiver, Penzias and Wilson started observing in the spring of 1964 at a frequency where they wouldn't expect much in the way of galactic radio waves. After this calibration they would then turn to a lower frequency to map the high-latitude gas they were interested in. But the telescope wouldn't co-operate. It was a great deal 'noisier' than they had been led to believe, or in Penzias's words: 'The antenna was considerably hotter than expected.'

A bad workman may blame his tools, but Penzias and Wilson were too competent for that. They persevered to find the 'fault'. One possibility was that two pigeons seen nesting in the horn might be responsible. The hapless birds were caught, posted to a nearby Bell Telephone Laboratories site and set free in their new home. Unfortunately, they flew straight back to the Holmdel horn and had to be discouraged 'by more decisive means'. But still the noise went on.

Early in 1965 Penzias and Wilson had reached a stage in their observations when they were able to dismantle their carefully calibrated telescope without harming their results. They meticulously cleared all the surfaces, removing large quantities of pigeon droppings (and an egg) in the process. Yet it all made hardly any difference. The noise was still there.

The last possibility was that the radio noise originated not in the telescope, but in the sky itself. If that was the case, it was spread out with a fantastic degree of uniformity. It showed no change with day, night or time of year, and it corresponded to a constant temperature of about 3 °C above absolute zero (minus 273 °C, the temperature at which all atomic motion stops). It was if the whole Universe was very, very slightly warm.

Penzias mentioned this puzzling result in a casual telephone call to a colleague one day. It rang a bell in the colleague's mind. He thought he'd recently come across a preprint of a paper by Jim Peebles at Princeton, which predicted that the Universe ought to have a background temperature of about 10 °C above absolute zero. Penzias contacted Robert Dicke, a senior colleague of Peebles at Princeton and a leading physics experimenter. Yes, it was true: if the Universe had begun as a hot, dense 'fireball', which exploded in a 'Big Bang' to produce our present expanding Universe, we should still expect some relic of that initial high temperature to pervade the whole of space. Peebles' calculations had shown that the background temperature expected today should be incredibly low, because it would have been cooled down by the expansion of the Universe. Coincidentally, a team led by Dicke had just started to search for the 'Background Radiation' with an antenna perched atop the Princeton Physics department roof. The final irony was that Peebles and Dicke found their calculations had been anticipated by almost twenty years anyway. In 1946, the Russian–American astrophysicist George Gamow – widely known for his popular science character Mr Tompkins in the books *Mr Tompkins in Wonderland* and *Mr Tompkins Explores the Atom* – had made a similar prediction in a paper that had long been forgotten.

Although Dicke and Peebles may have felt that here, at last, was proof that the Universe began in a hot 'Big Bang', Penzias and Wilson themselves weren't so certain. When the two groups published their findings later in 1965 as companion letters in *The Astrophysical Journal*, Penzias and Wilson simply called their paper 'A Measurement of Excess Antenna Temperature at 4080 Mc/s', and referred readers to the 'theory' paper for interpretation.

Despite the understandable caution of Penzias and Wilson, it's seldom that the astronomical community has embraced a new finding quite so quickly or enthusiastically. There's very little else the radiation *could* be, other than the cooled-down relic of an earlier hot phase of the Universe. Of course, astronomers had to check its energy distribution, with different telescopes and at different frequencies, to make sure that it wasn't something closer to home – such as a huge nearby dust cloud – but the answer was the same, every time. Instead of being at absolute zero, the Universe is bathed in radiation which raises its temperature to 2.7 degrees above this frozen limit. We say that the background temperature is 2.7 degrees Kelvin (or 2.7 K: a Kelvin degree is a Centigrade degree measured from absolute zero rather than 0 °C). Rounding off, people often refer to it as the 'three degree (or "microwave") background'.

While this discovery was generally welcomed enthusiastically, it left some astronomers disappointed – even sad. At the time, a good many scientists (and philosophers) went along with an alternative theory about the nature of the Universe, first put forward by Fred Hoyle, Hermann Bondi and Tommy Gold in 1948. It was a beautiful, elegant theory which dispensed with messy notions of 'beginnings' and 'endings' by a process of 'continuous creation', in which atoms were slowly generated by 'negative energy' fields in space. The point made forcibly by the *Steady-State Theory* is that the Universe is changeless. Although it is expanding, galaxies are continuously created to fill the gaps left. And wherever you look in the Universe – no matter how far away in space, or back in time – it will always look the same.

Britain's present Astronomer Royal, Sir Martin Ryle, challenged the last point in the late 1950s with evidence that distant radio sources seemed closer together than they should. Perhaps the Universe had evolved from a denser state? But his statistics didn't convince the Steady-State stalwarts. Penzias and Wilson's discovery, however, did. Leading theorist Dennis Sciama is typical when he talks of his 'sadness at losing a beautiful theory'; but, like so many, he was won over by the apparently incontrovertible evidence of the background radiation.

Today, the vast majority of astronomers agree that our Universe began some 15,000 million years ago in a hot 'Big Bang'. The evidence is doubly persuasive. The microwave background is just one part of it; we also have to take into account the remarkable coincidence between the birthdates of the oldest stars in galaxies and the result of extrapolating backwards the galaxies' present outward flight to a point in the past when they – or rather, their raw materials – were all together. Both work out to a figure of between 15,000 million to 18,000 million years in the past.

Astronomers are adept at the technique of backward crystal-gazing, for the finite speed of light and all other radiations means that everything they see is of the past. They use this trick to the limit when it comes to describing the origin of our Universe. But they play the game with strict rules rather than speculation. They take all the observed facts – the expansion of the Universe, the structure of the Universe, the age of the galaxies, the temperature of background radiation, the balance of chemical elements, the evolutionary trends they believe they can detect – and then, using the rigid code of the laws of physics, they turn the hands of the clock back in time towards zero. Even so, the job of a cosmologist (someone who studies the structure and nature of the Universe) is not easy. The highly-respected Cambridge cosmologist Martin Rees speaks from the heart when he says: 'Cosmology is a peculiar science. It is by definition the study of a unique object and a unique event ... No biologist would formulate general ideas of animal behaviour after observing just one rat, which might have its own peculiarities. Yet cosmologists are shown one rat, doing just one thing, and are asked to explain the behaviour of rats on the basis of that observation.' But Rees maintains that there's some justification for prediction because 'the Universe, in its large-scale structure seems simpler – smoother and more symmetrical – than we had any right to expect.'

Because of the precision of the laws of physics – particularly on the scale of the very small – we can peer back into the history of our Universe to a point only 10^{-38} seconds (that's 0.000, 000, 000, 000, 000, 000, 000, 000, 000, 000, 000, 000, 01 seconds!) after it was born. And it appears that its entire future fate was decided within the first hundred *seconds* of its existence, right from its present balance of elements through to whether it was destined to contain any matter at all. Ironically, the period after its first few minutes until its thousand-millionth birthday is shrouded in mystery, because there are so few theoretical or observational clues to guide us. Martin Rees, however, cautions that we dare not be too complacent. 'Our present satisfaction [with the Big Bang theory] may reflect the paucity of the data rather than the excellence of the theory.'

At the instant of the Big Bang, both space *and* time were set to zero. There is no 'before the Big Bang'. That's because time and space are inextricably linked to form a 'space-time continuum', as Albert Einstein demonstrated in his theory of relativity. Without space, there can be no time.

But it's not for astronomers to say what caused the Big Bang. If they extrapolate back towards the instant of creation, they must push the laws of physics beyond their legitimate limit – and that's not in the rules of the game. All they can say is that the Big Bang happened, setting into motion our expanding Universe; and that when it happened, the temperature and density of the Universe must have had values approaching infinity, that is almost infinitely high. By the moment only 10^{-38} seconds after expansion began, the density of the young Universe had dropped to a point at which we can proceed – with great caution – to gauge conditions there by using our knowledge of the laws of physics. We have no guarantee that the same laws were valid so long ago, or if conditions in the raging inferno of the Big Bang may have given rise to effects we can't even guess at. But we can take some comfort from the fact that these same laws have held up in environments as extreme as the surroundings of a black hole and in locations as remote as the farthest quasar. Besides, we have no other recourse open to us – apart from idle speculation.

Running the film backwards to the limit, we can begin to get an idea of the unbelievable extremes that our young Universe had to live through. At this point, only fractions of a trillionth of a second after creation, physicists can tell us that the temperature of the Universe was a staggering 100,000, 000,000,000,000,000,000,000,000,000 °C, and that its density was an even more mind-boggling 1,000, 000, 000, 000, 000, 000, 000, 000, 000, 000, 000, 000, 000, 000, 000, 000, 000, 000, 000, 000, 000, 000, 000, 000, 000, 000, 000, 000, 000, 000,000,000 times the density of water. Matter, as we know it today, couldn't possibly have existed. Instead, the Universe was a concentrated soup of searing gamma rays, chock-a-block with the ultimate subatomic particles which were later to combine to make up atoms of matter – and antimatter.

Every particle of matter has its 'antimatter' equivalent, opposite in every way (spin, charge etc) except mass. The Universe today is undoubtedly made of matter, and antiparticles are extremely rare – they're sometimes detected in particle physics experiments on powerful accelerators – but once discovered, they don't last very long. Whenever matter and antimatter meet, they spontaneously annihilate each other.

The very fact that we are here at all appears to have been decided at this stage of the Universe's existence. Fractions of a second after the Big Bang, there must have been equal numbers of particles and antiparticles around. Although they were annihilating all the time, new particle-antiparticle pairs were being spontaneously and continuously created to replace them. The agent was the energy of the Universe itself – the powerful radiation field filling all of space. Just as matter undergoing nuclear fusion can create energy, so powerful energy fields can generate particle-antiparticle pairs.

But by the time the Universe was a hundredth of a second old, expansion had caused its temperature to fall to a 'mere' 100,000 million °C, too cool to create matter and anti-matter any longer. All the remaining particles and antiparticles should then have totally annihilated each other for once and for all, leaving the Universe as an empty bath of searing radiation. Had things happened like this, the presence of matter (and antimatter) in the Universe would be just a long-forgotten aberration; a freak event which took place briefly in the trauma of the Big Bang. Today, the Universe would be an ever-expanding, steadily-cooling emptiness.

But it seems that our young Universe was not quite as symmetrical as it should have been. The radiation field must have generated a very slight excess of matter over antimatter: 100,000,000 particles to every 99,999,999 antiparticles. Those figures, so tantalisingly similar, mean that one matter particle in every hundred million survived. This barely perceptible inbalance was to lead to our present Universe of stars, planets and galaxies – and our own existence in it.

There's no doubt, though, that radiation must have had the upper hand during the first million years of our Universe's life. This was especially true during the initial few seconds after the Big Bang; but to find out how greatly the radiation dominated, we have to view it from a slightly different perspective. Although until now we have looked on light (and all other radiations) as being the vibrations of an electromagnetic wave, we can also regard it as a stream of particles. These massless *photons* travel at the speed of light, and experiments show that they can behave purely as particles – as, for instance, in the photoelectric effect – or in the more conventional manner as waves. Counting the number of photons in a given volume can tell of the intensity of a radiation field.

About two seconds after the Big Bang, there were around a thousand million photons to every particle of matter. Radiation certainly called the tune. What little matter there was, still in a sub-atomic, electrically charged state, was subject to all the vicissitudes of the radiation's all-powerful and enveloping electromagnetic field. But matter had an ally in the expansion of the Universe, which continually cooled and diluted the radiation bath.

By the end of the first 200 seconds – about $3\frac{1}{2}$ minutes after the Big Bang – matter had scored its first decisive victory over radiation. Encouraged by the weakening field, some of the subatomic particles had managed to knit themselves together to form protons – positively-charged heavier particles which today make up the nuclei of atoms. Single protons, as we found out in Chapter 7, are the nuclei of hydrogen atoms.

Other subatomic particles locked together in a different way to produce a particle very similar to the proton, but without an electric charge: the neutron. And the tiny, negatively-charged electrons were there too, maintaining their identity as single particles.

Near the end of this period of consolidation, the radiation field had weakened just enough to allow some of the protons to pair up with neutrons. This proton-neutron combination makes up the nucleus of a deuterium atom – the 'heavy hydrogen' we use in today's H-bombs – but it's an unstable alliance. It can be strengthened only if two deuterium nuclei pair up with one another, which is just what must have happened at the end of the period. But this double-pairing destroys deuterium's identity: the two-proton, two-neutron combination formed is instead the nucleus of a helium atom.

After $3\frac{1}{2}$ minutes, the Universe had formed its basic building blocks: the nuclei of the atoms hydrogen and helium. It could go no further than this, because the powerful radiation field would have blasted apart any more ambitious attempts at element building. But for the moment, it was enough.

One of the outstanding successes of the Big Bang theory lies in its ability to predict just how much hydrogen and helium were formed during this period of matter's first bid for power. Working through the physics, we find that three-quarters of the protons should have remained single, while a quarter will have combined into double proton-neutron pairs. And that's exactly the proportion of hydrogen to helium that we observe today, in the oldest unchanged stars.

Even after this small victory, matter's battle was by no means over. Both the electrons and the newly-formed positive nuclei were overwhelmed by the strong surrounding electromagnetic field. The Universe was an opaque, seething battleground in which matter and radiation constantly parried and clashed. And so it went on until the Universe was a million years old.

By now, the temperature had fallen to only 3000 °C (5400 °F), and the radiation field which had previously kept the nuclei and the electrons apart had weakened to a point where it was no longer able to do so. The electrons could at last combine with the nuclei to make atoms of hydrogen and helium. And at this point, radiation lost the battle. Particle-for-particle, it still dominated matter by 1000 million to one. But it could no longer have a serious effect on it, for atoms, being electrically neutral, are essentially unaffected by electromagnetic fields. At this point, too, space became transparent as matter and radiation ceased to interact with each other.

Still the radiation remained, cooling and weakening all the time as the Universe continued to expand, until it reached the stage in which we find it today; a barely-detectable presence which keeps space from freezing to absolute zero. It's this tell-tale background radiation, produced when the Universe was last opaque, that Penzias and Wilson first picked up with their horn receiver at Holmdel. They were looking back to the furthest point in time that anyone can – to our Universe's millionth birthday.

The stage would appear to be set, then, for the next major event in the history of the Universe: the birth of the stars, planets and galaxies. There were abundant raw materials, and conditions were improving all the time; and the galaxies we see today must have been formed sometime between this era of 'recombination' and the present day. But when? For all their near-certainty over the earliest phases of our Universe's existence, astronomers are noticeably reluctant to specify exactly when the galaxies were formed. Their caution is quite understandable, because the galaxies have left very few clues as to when, and how, they were born. But it is certainly ironic that we believe we know more about the Universe's first fractions of a second than we do of the next few thousand million years of its life.

The field of galaxy formation is one of the most uncertain areas in astronomy today. No one has ever seen a galaxy being born. And it's hard to be sure just what a young galaxy should look like – although astronomers have tried to run back the clock on today's galaxies to find out. Even so, scientists hunting for 'protogalaxies' have the odds stacked high against them when they can't be sure how their quarry is going to behave. All they can say is that galaxies came into being sometime between 100 million years and a few thousand million years after the Big Bang – before that, it was too hot, and by the end of the period, young but definitely formed galaxies (quasars) are visible anyway.

In the absence of firm evidence from the past, astronomers have to be content with a broad-brush picture of galaxy formation which is at least

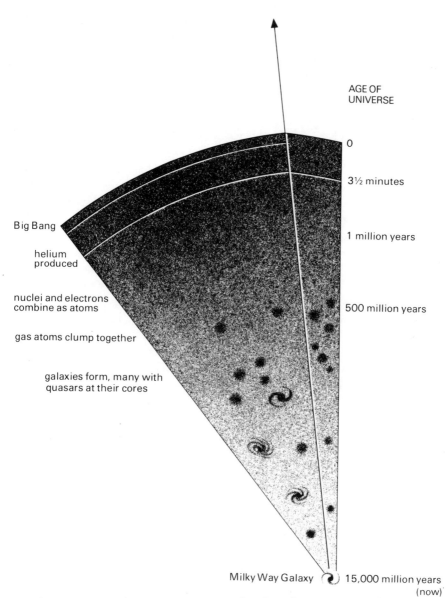

looking outwards in space,
backwards in time

AGE OF
UNIVERSE

0

3½ minutes

1 million years

500 million years

Big Bang

helium
produced

nuclei and electrons
combine as atoms

gas atoms clump together

galaxies form, many with
quasars at their cores

Milky Way Galaxy   15,000 million years
(now)

23 A 'CROSS-SECTION' OF THE UNIVERSE *Light takes a finite time to travel across space,
so as astronomers look farther out into space, they are also looking backwards in time.
This is a 'cross-section' of the Universe, centred on our Milky Way Galaxy. (The diagram
does not mean that our Galaxy is 'at the centre of the Universe': the Universe has no
centre, and astronomers in any other galaxy could just as validly draw a sphere centred on
their own galaxy.) Whichever way we look, at a distance of 15,000 million light years we
are seeing back to the instant of the Big Bang, which occurred 15,000 million years ago
and created our Universe, space and time itself. The events soon after the Big Bang lie
slightly closer to us in both space and time, as shown here.*

consistent with the way galaxies are today. We have a good deal of evidence that galaxies must have somehow collapsed from gas clouds, and so astronomers conclude that there was a time in the past when the hydrogen and helium atoms filling the Universe began to clump together. For this to happen, conditions must have been cool enough for the atoms to slow down sufficiently to come under the influence of each other's gravity. Although still carried relentlessly apart by the expansion of the Universe, a small clumping of atoms in one place would rapidly attract others to the spot. Clouds would quickly form, still expanding outwards in step with the Universe.

In the next stage, these clouds must begin to collapse. If there are enough atoms in a cloud to overcome the expansion of the Universe by their own gravitational inpull, collapse will be certain and rapid. And by way of confirmation, theorists calculate that the preferred three sizes for collapsing clouds are indeed the same as the characteristic masses for the largest groupings of matter we find today: clusters of galaxies, single galaxies and globular clusters. But did small clouds bond together to make up galaxies and clusters? Or did collapse occur simultaneously on all three scales?

Whatever the details of this process, the result is millions of protogalaxies – each 'a system of gas clouds racing in orbit and giving birth to stars when the clouds collide', in the words of protogalaxy hunters David Meier and Rashid Sunyaev. And because of the way that all gas clouds collapse, scientists can begin to predict what a protogalaxy should look like. Collapse always occurs fastest towards the centre of a gas cloud. In the case of a protogalaxy, this would trigger a massive burst of star formation at the core, causing it to shine hundreds of times brighter than its surroundings. In appearance, a protogalaxy would be difficult to distinguish from a quasar, but a spectroscope could quickly spot the difference: in a protogalaxy's spectrum there would be tell-tale lines coming from young stars.

To date, no protogalaxies have been identified, but scientists aren't too despondent. Although intrinsically very luminous, protogalaxies are probably just beyond the range of present-day telescopes. But by the mid-1980s we should have a better chance because, as David Meier predicts 'The Space Telescope may be able to detect all primaeval galaxies except those with the highest redshift' (see below).

At this point we could start to bring the story of galaxy formation to a close – except that we need to explain just how the same initial conditions could have given rise to such a wide range in galaxy types. As we found in Chapters 14 and 15, there is a world of difference between a dwarf irregular galaxy and an exploding cosmic spectacle like Centaurus A. Did all galaxies really start off life in the same way?

Groping in the dark, astronomers have come up with the following scenario. Elliptical galaxies form from gas clouds (or a collection of clouds) with little or no rotation. Collapse is straightforward, and stars are born rapidly, using up all the available gas in a relatively short space of time. If the protogalaxy happens to be a very big one, collapse at the centre may lead to the formation of a massive black hole, which, as we saw in Chapter 15, can give birth to a quasar. The quasar's continuing violence at the galaxy's centre can scour the young galaxy of any remaining star-making gas. And so the giant ellipticals we see around us today are rather disturbed, anaemic creatures whose stars were among the first in the Universe to be born.

We can't apply the same reasoning to the tiny dwarf ellipticals. Their central density could never have risen high enough to produce a galaxy-scouring black hole and accretion disc; and yet, they are also pathetically short of gas. Maybe it's no coincidence that we only find them close to larger galaxies whose tidal pull can slowly drag gas out of them.

It's just as hard to explain why irregular and spiral galaxies are blessed with an abundance of gas. As we saw in Chapter 14, some of the smallest irregular galaxies are little more than giant gas clouds which take the process of star formation at a snail's pace. A few may have only recently got around to making stars. Perhaps the intergalactic gas clouds haven't even got started. It's by observing systems like this that astronomers hope to get clues about stellar birthrates and galaxy formation in general.

When it comes to large irregulars and spirals, the difference appears to be one of mass. Both types rotate, but only galaxies with more than about 10,000 million stars are able to 'grow' spiral arms. However, it seems that it's their relatively rapid rotation, and comparatively medium size, that has left them rich in gas and capable of prolific star formation even today. Astronomers think that spirals (and probably irregulars) began life as gas clouds which collapsed symmetrically – and the spherical shape of our galactic halo is good evidence of that. But unlike the proto-elliptical galaxies, these collapsing protogalaxies were very slowly spinning, given a slight tug, perhaps, by the close passage of a near-neighbour.

Any slight spin is amplified when collapse gets under way. A proto-spiral galaxy would spin ever more swiftly as it shrank in size, developing flattened poles and a prominent bulge about its equator. Computer simulations reveal that the next stage is one of sudden chaos as the bulge of the spinning protogalaxy collapses to form a wide disc around the already-formed core. The disc rebounds several times before head-on collisions of gas clouds inside rapidly damp its vibrations. There's a burst of star formation in this newly-created disc; but now the gas is spread much more thinly than before. Unless the galaxy is able to drive away the gas by means of Seyfert outbursts in its core – which are anyway far less powerful than quasar or radio galaxy eruptions – it is assured of a large supply of low-density gas which will slowly form into stars over a very long timescale.

Astronomers are painfully aware that their ideas on galaxy formation could be very wide of the mark. It's an almost impossible task to wind back the involved, complicated and interactive history of a collection of millions of stars, without even being sure what conditions were like when the galaxy first started to collapse. Before this decade is out, however, the unrivalled light-grasp of the Shuttle-launched Space Telescope should tell us more of these early days as it picks out distant protogalaxies.

Before we come back to the present, there are just a few other dates in the history of the Universe which deserve mention. Unlike the previous dates, which marked important events in the Universe's life, these are far more parochial: they are only of concern to us. Of especial interest is a date about 10,400 million years after our Universe's birth. It was then that yet another massive star in an undistinguished galaxy exploded, showering the space around it with its processed star-stuff, and causing nearby gas clouds to collapse almost instantly. As a result, less than a million years later (4600 million years ago) a small yellow star and its nine planets came into being. And almost 15,000 million years after the Big Bang – only a million years in the past – a hairless bipedal creature of remarkable dexterity arose on one of the smaller planets of the yellow star. He and his race rapidly came to dominate the planet, just as the dinosaurs and woolly mammoths had done before him. He called his planet 'Earth', his star 'the Sun' and his galaxy 'The Milky Way'.

And now he has started to look back beyond his own only-too-recent origins to seek answers to the question of how everything began. Until quite recently, he wasn't doing too well, as Nobel prize-winning physicist Steven Weinberg relates: 'I remember that during the time that I was a student and then began my own research (on other problems) in the 1950s, the study of the early universe was widely regarded as not the sort of the thing to which a respectable scientist would devote his time.' But dramatic strides in elementary particle physics, coupled with the serendipitous discovery of the microwave background radiation, allow today's astronomers a hazy glimpse into the raging inferno of the Big Bang itself. Perhaps we are beginning to read the past with some small degree of certainty. Will we ever be able to say the same about the future?

— 17 —
The End of Knowledge

We are living at a time when the Universe is in the prime of life. Its early life was a period of unchecked violence, as the Universe sprang into existence at the dawn of time itself in the mighty inferno of the Big Bang. Even after galaxies of stars had formed out of the original swirling gas clouds, many a galaxy was initially dominated by the ferocity of a quasar at its core. Now most galaxies, including the Milky Way, are busy bustling enterprises, where stars are born and die – sometimes interrupting the harmony in a supernova outburst – and new generations of stars follow them. Planets wheel undisturbed around the stars, and living beings on these planets can enjoy the Universe in its heyday.

But the Universe cannot remain in its prime for ever; like all things, it will pass its best and fade away. Galaxies are mortal. They will reach senility and long, cold lingering death as their last stars burn out.

With their powerful telescopes of all kinds, and their equally-powerful theories, astronomers have been able to examine the birth-pangs of our Universe, and to trace its development up to the present day. Those theories can also push our knowledge ahead in time, to let us predict the ultimate fate of the Universe. These results are not pure conjecture. We know that the laws of science work everywhere, from under the microscope in a laboratory to the farthest galaxies and quasars known; and they have worked from an instant after the Big Bang up to the present day. We can confidently use these same laws, and the theories that incorporate them, to calculate what is in store for the Universe. An astronomer can predict the state of the Universe in a million million million million years' time with far more confidence than a meteorologist can foretell next week's weather!

In the wake of the Big Bang, the Universe was set expanding. It is continuing to expand today: as we've seen in previous chapters, clusters of galaxies are all moving apart from one another, and in the near future the momentum of the giant clusters of galaxies must keep them moving apart. But will the Universe keep expanding for ever?

This is really the key question – the question that must be answered before we can safely predict the far future. There is one force that always opposes the expansion of the Universe, and that is the force of gravity between the clusters of galaxies. Gravity always attracts one body to another, and the attraction between the clusters of galaxies tries to slow down their separation. It is a tussle between the momentum of the expanding Universe and the long-range force of gravity. If the galaxies' momentum is sufficient, the Universe will continue on its course of expansion for all eternity. But if gravity wins the upper hand, its attractive force will slow down the speeding clusters of galaxies and bring them to a standstill. And then, gravity will draw them all together again. The Universe will shrink, until all its matter runs together in one tremendous implosion. As the exact opposite of the Big Bang which gave life to the Universe, this possible fiery infall is called the Big Crunch.

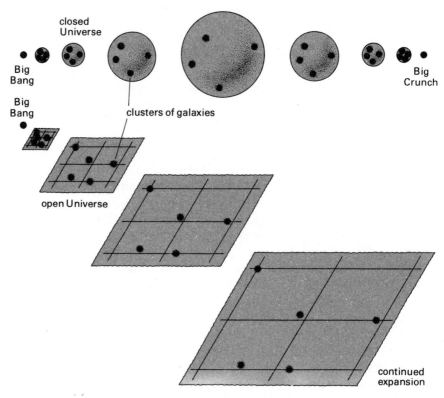

24 POSSIBLE FUTURE FATES FOR THE UNIVERSE *Cosmologists visualise our Universe as a 'surface' in another dimension: here the Universe is represented as a two-dimensional expanding surface like a very thin sheet of rubber. According to cosmological theories based on Einstein's general theory of relativity, the Universe can take one of the two forms shown here. If the Universe is bent back on itself (top) like the surface of a balloon, then it cannot continue expanding forever; eventually it will reach a maximum size, and will then contract to end in a 'Big Crunch'. Alternatively, the Universe might be 'open', like an infinitely large piece of rubber sheet, and in this case it will continue to expand forever. Present astronomical evidence favours the latter alternative, so the Universe will always expand even after the stars and galaxies die.*

Supposing that gravity does win at some time in the distant future, we can calculate a timetable for the catastrophic count-down to the Big Crunch. American scientist Freeman Dyson paints the following picture. A thousand million years before the Big Crunch, the clusters of galaxies, speeding together, have closed up the empty space between them and begin to merge together. By a hundred million years before, the space between individual galaxies has closed up, and all galaxies merge together: the entire Universe is now filled with stars at roughly the same spacing as the stars in our Galaxy.

To any of our descendants looking from their world (for the Sun and Earth would long since have died), the night sky would now be ablaze with stars. Night would be as bright as day. But this pleasant state of affairs would not last long. Space is still shrinking, and now that the galaxies have merged, gravity is pulling the stars closer and closer together. At a time about 100,000 years before the Big Crunch, the stars are so close that the night sky is not just blindingly bright, but searingly hot. It must be the end of any living beings on any planet in the Universe. Under the incandescent skies, oceans will boil away; before too long, stars will pass so close that their heat will melt the solid rocks, and the molten rocks will boil away into space. By a thousand years before the Universe destroys itself in the Big Crunch, the stars will be colliding with each other in their hundreds. Each resulting mélange collapses into a black hole. And in the final act, these black holes merge to become one; within it, all the matter of the one-time Universe is shrunk to a tiny, infinitely compressed singularity.

Some astronomers believe that the Big Crunch, if it happens, will instantly re-explode as another Big Bang, and usher in the birth of a new universe. But even if this is the case, there is no ultimate hope for life in a Universe where gravity wins control: nothing can survive the holocaust leading up to the Big Crunch.

It doesn't look as though the Universe will end up in a Big Crunch, though. The latest evidence indicates that gravity will always play second fiddle; that the galaxies' momentum will win, and keep the Universe expanding and expanding for ever into the future.

The deciding factor is simply the average density of matter in the Universe. Imagine smearing out the matter in galaxies and between galaxies, until it filled space absolutely evenly. Since most of space is virtually empty, the average density is very low. If the amount of matter in a cubic light year of space is about the same as the mass of our Moon, or more, then gravity will have the upper hand and we are destined for a Big Crunch; if less then we are destined to perpetual expansion. When astronomers work out this calculation, mentally 'smearing out' all the matter, they find there's only one-thirtieth of the matter needed to lead to a victory for gravity.

But there is one complication. Most astronomers now believe that the Universe contains a huge amount of invisible matter, far greater than the amount of matter making up the stars and gas that we see in galaxies (Chapter 14). If there's enough 'missing mass', it may be sufficient to give gravity the upper hand. We can't detect the missing mass directly, but astronomers have deduced its presence by its gravitational effect on stars and galaxies in the Universe today. In the past few years, several teams of astronomers have tried to determine just how much missing mass there is, and they conclude that it outweighs the matter we can see by about ten-to-one. So, although there's probably a lot of invisible matter, of some unknown kind, around and between the galaxies we see, it still falls short of the amount needed to give gravity mastery over the Universe. The Universe will indeed expand for ever.

As all the clusters of galaxies continue to speed apart, our distant ancestors will see other galaxies recede further and further into the distance. Only the few dozen members of our Local Group of Galaxies – including the Andromeda Galaxy – will remain to keep us company. Five thousand million years from now, the Sun will die – and in its final bloated swelling as a red giant, it will destroy the Earth. But humans should, by then, have moved on to the planets of younger stars.

Our Galaxy cannot keep on spawning stars indefinitely, though. Stars condense from the general interstellar gas; and although they return some of this gas on their death beds, most of it is locked up in dead star corpses – white dwarfs and neutron stars. As time goes by, more of the Galaxy's matter will be in the shape of these dark, dead relics, and less will be available as gas for creating new generations of stars. Eventually, the last tatters of gas in the Galaxy will condense into one last generation of stars. These last stars will barely illuminate the vast star-system, consisting mainly of the cold shrunken corpses of previous generations. And then they, too, will die, to leave the Milky Way a dark and lifeless stage, the final curtain in a cosmic Hamlet.

Science can however take us further, to a cosmic curtain-call in which the corpses take on a new – and different – lease of life. The final generation of stars will die about a million million years from now. If we wait a further million million million years, then we'll find that something new has happened. The dark white dwarfs and neutron stars orbiting in the now defunct Galaxy have gradually affected each other's paths so that some are thrown out of the Galaxy altogether, while the others – to compensate – have crowded down towards the centre. Here they eventually coalesce. The result is a huge black hole containing much of the Galaxy's original matter, and drifting away outwards a halo of escaping white dwarfs and neutron stars.

And we still have not reached the end of the road. As we saw in Chapter 10, black holes are not forever. They gradually 'evaporate' into subatomic particles and radiation, ending up in a spectacular 'explosion'. The black hole which was once our Galaxy will eventually explode in a flood of particles and radiation, some one thousand million million million million

million million million million million million million million years from now.

At this point in the future, even the scientists' crystal ball begins to go hazy. Our Universe will then be a very empty, dark place. Here and there, we would come across an odd subatomic particle, the only relic of the once-brilliant galaxies, and occasionally we'd pick up a weak radio signal, the final diluted radiation from the fury of the Big Bang in which the Universe began.

Victorian scientists likened the Universe to a clock; once wound up tightly (at the instant we nowadays call the Big Bang) it is gradually running down, losing its available energy, until eventually its wheels stop turning – all activity in the Universe ceases. Some modern scientists however have suggested that something can give a little kick to the clock-spring, can inject motion and vigour into a Universe that should by rights be dead and still. That 'something' consists of the singularities which reside at the centres of black holes. After the black holes have exploded, these singularities will be left, exposed 'naked' to the Universe at large. As we saw in Chapter 10, a singularity can defy the laws of science. It can generate energy, light and heat from nothing; it can spontaneously produce matter, too, perhaps spewing out a planet, a bottle of vintage wine, or something we can't even imagine. Perhaps the Universe will ultimately enter a new stage of existence. Instead of being merely the slowly-cooling embers of the Big Bang – as it is today – the Universe will be constantly renewing itself, through the unpredictable antics of a host of singularities. If this is so, then there may be a chance for life to survive, and to pass into a new kind of existence.

This is speculation at the borders of known science. This future lies far, far away; and even if we can predict how the inanimate matter of the Universe may behave, it's not so easy to predict the future of man himself. Intelligence may turn out to be a powerful new force in the course of our Universe's future evolution.

Only four hundred years ago, most men believed that the Earth was the centre of the Universe, and the stars merely pin-pricks in a large surrounding dome. Even forty years ago, we had no idea of when or how the Universe was born. Twenty-five years ago, Sputnik 1 opened the way to space. Compared to the age of the Universe, or the age of the solar system – both measured in thousands of millions of years – these are staggeringly short times. In the twenty-five years of the 'space age', our progress has been even more astounding. Men have walked on another world, the Moon, and brought a third of a tonne of it home with them; robot spacecraft have grubbed in the deserts of Mars, survived Venus's red-hot crushing acid bath, photographed all the planets out to Saturn, and have even taken paths which will lead them away from our solar system to the realms of the stars. Already, men have drawn up plans for dismantling planets, rehousing most of humanity in space, travelling in person to the stars. Given the time, what more could we achieve: the colonisation of the Galaxy, the control of pulsars and black holes, the taming of the mysterious singularities which alone can prevent the winding-down of the Universe? We have 5000 million years until the Sun dies; almost a million million years until the final stars in our Galaxy flicker out. By that time, we may not be merely helpless bystanders, impotently watching the wreck of the great ship we call the Universe; the factor that present-day astronomers have left out of their pessimistic predictions may, ironically enough, be ourselves.

Some of today's astronomers do recognise that Man has an unusual relationship with the Universe around. As we have come to understand more about the construction and workings of the Universe, it's become more and more evident that our Universe is ideally suited to the development of intelligent beings like ourselves.

The long line of development that's led to our appearance on Earth has required a whole string of 'coincidences', working hand-in-glove. Some are fairly obvious, others fairly abstruse; and they range from the borderlines of philosophy to the intricacies of nuclear physics. Take the fact that the Universe is an orderly place, for a start. Once a pattern is established, it won't change unless something else affects it, and then it will alter in a predictable way, according to the rules of the game – the laws of Nature, in more scientific terms. So, for example, the Earth has circled the Sun in more or less the same orbit for the four thousand million years that life has taken to evolve to human beings. That's in the rules of the game. But we could equally well imagine a Universe with different rules – or without rules – where the Earth could have jumped frivolously about, at one moment jumping in to Mercury's sun-parched position and at another leaping out to the frigid cold of interstellar space. Clearly life would be impossible in such a disorganised type of Universe.

As well as being a well-ordered place, our Universe seems very careful in its choice of rules – and always seems to have chosen laws of Nature which are favourable to the development of intelligent life. British cosmologist Martin Rees puts it this way. 'All the features of the everyday world and the astronomical scene are essentially determined by a few basic physical laws and constants, such as the masses of the elementary particles, and the relative strengths of the basic forces that operate between them. In many cases, a rather delicate balance seems to prevail.'

Nuclear reactions, for example, involve a balance of the 'strong' nuclear force and the electric force, the former attracting two subatomic particles together, the latter repelling them. This balance is on a razor-thin knife-edge as far as we are concerned. If the nuclear force were just a little bit weaker, then subatomic particles could not join together to form the nuclei of atoms. The only type of atom would be hydrogen (which has a single proton in its nucleus), and without the other elements, like carbon, oxygen and nitrogen, life just could not exist. On the other hand, if the nuclear force were just a little stronger, then all the original hydrogen of the Universe would have turned to helium in the Big Bang, and there would be no hydrogen about today. And hydrogen is essential to us in two ways: it is a vital component of the chemicals which make up living cells, and it is the 'fuel' which keeps the Sun shining – in fact, hydrogen is the only nuclear fuel which can keep a star shining for the thousands of millions of years that intelligent life needs to evolve on an accompanying planet.

There are similar, crucial, balances in the strengths of the other forces, the 'weak' nuclear force and the force of gravity. On top of that, the actual masses of the subatomic particles are important too. If the proton were not 1836 times heavier than the orbiting electrons in an atom, then chemistry would be different – and in particular, the complex molecules essential to life (like proteins and DNA) would not be stable.

There are far too many 'coincidences' to be mere chance. We must conclude that we are only here because the Universe has certain, very specific, rules built into it. Why? One could invoke a God who has set up the Universe this way just for us to inherit. Many scientists, however, think that this is shirking a proper investigation: if we had meekly accepted that lightning was caused by Thor throwing thunderbolts, we would now know very little about electricity, and we would never have had the understanding to protect buildings (and lives) with lightning conductors.

Most scientists who've looked at these coincidences have come to a different conclusion, the 'anthropic principle'. Stated in its boldest terms, the anthropic principle asserts that the Universe is the way that it is *because* we are here to observe it. That may sound like putting a very large cart before a very small horse, but one way of comprehending the argument is to imagine that Nature has created a huge number of different universes. All of them are built to work according to different rules – and some may have no rules at all.

The universes with no rules end up in complete chaos: there's no chance of intelligent life arising there. And most of the remaining universes will have rules that don't balance well enough to allow the development of long-lived stars, stable chemical elements, or the complicated compounds that make up life. As Martin Rees explains 'most of the universes are still-born, in that nothing interesting ever happens in them. In only some of them can complex structures evolve into any kind of conscious entity.' Our Universe contains conscious entities – namely us – so it must be one of that small fraction of universes which are well-ordered and well-balanced. Our presence here is thus intimately linked with the orderliness and knife-edge balance we find amongst the laws which govern our Universe: if we want to argue which is *because* of which, we are in a chicken-and-egg rather than a cart-and-horse situation.

The anthropic principle underlines our interrelation with the Universe – and the only possible alternative, a Universe designed by God for man, makes the same point even more strongly. The Universe is part of every one of us, and at many levels: from the mere physical fact that our bodies are composed of stardust, to our ultimate heritage, the use of our intellect to conquer space. Given time, mankind could spread over our Galaxy, and make the Milky Way his home, just as our ancestors left their homelands to take over this planet on which we live. Less than five hundred years ago, Christopher Columbus set off out across the wastes of the Atlantic to find the Indies, and discovered a New World; in 500 years' time, men will probably be on their way to the stars: in 500,000 years, a new Columbus may be setting sail on the two million light year void between the Milky Way and the Andromeda Galaxy.

We have only just begun to dabble our toes in the great oceans of space. In Chapter 1, we saw these seas as hostile: space is cold, it is airless, and filled with dangerous radiations; further afield it bristles with predators – pulsars, supernovae, black holes. Yet this is only a first impression. We make comparisons with the lush, safe haven of planet Earth, and the life we lead here, in the late twentieth century. But space need hold no fears, if we can take our habitat with us. And we are already a long way along that road. Civilised western man lives in an artificial environment now, with light and heat at the touch of a switch, and virtually instant food and medications available; in a centrally-heated hotel bedroom, with television, auto-bar and room-service, we can make a winter in Montreal more comfortable than Robinson Crusoe's tropical paradise.

Earth will, however, always be our spiritual home. The Apollo astronauts, winging to and from the Moon, realised that. After returning from Man's first trip around the Moon in Christmas 1968, William Anders said 'the Earth looked so small, insignificant and fragile. It seemed such a shame that people aren't motivated to live together better, conserve the planet [and] stop air pollution.' But Man is not content to stay on this one planet with a whole wide unexplored Universe before him. As the great Russian rocket pioneer Konstantin Tsiolkovsky wrote at the beginning of this century 'the Earth is the cradle of the mind, but you cannot live in a cradle for ever.'

And after reviewing some of the 'coincidences' in the basic structure of the Universe, Freeman Dyson asserts: 'I believe the Universe is friendly. I see no reason to suppose that the cosmic accidents that provide so abundantly for our welfare here on the Earth will not do the same for us wherever else in the Universe we choose to go.' The more we discover about the Universe, the more we comprehend that we are an integral part: the Universe is not something 'out there', divorced from the rat-race scrabblings of Man on this planet. Man is as much a creature of the Universe that made him as he is of the planet that has nurtured him. The Universe is part of our existence, as relevant as the air we breathe, and as close and unspoken as the bonds of family kinship: it is *our* Universe.

INDEX